Lecture Notes in Physics

New Series m: Monographs

The Editorial Policy for Monographs

The series Lecture Notes in Physics reports new developments in physical research and teaching - quickly, informally, and at a high level. The type of material considered for publication in the New Series m includes monographs presenting original research or new angles in a classical field. The timeliness of a manuscript is more important than its form, which may be preliminary or tentative. Manuscripts should be reasonably self-contained. They will often present not only results of the author(s) but also related work by other people and will provide sufficient motivation, examples, and applications.

The manuscripts or a detailed description thereof should be submitted either to one of the series editors or to the managing editor. The proposal is then carefully refereed. A final decision concerning publication can often only be made on the basis of the complete manuscript, but otherwise the editors will try to make a preliminary decision as definite as they can on the basis of the available information.

Manuscripts should be no less than 100 and preferably no more than 400 pages in length. Final manuscripts should preferably be in English, or possibly in French or German. They should include a table of contents and an informative introduction accessible also to readers not particularly familiar with the topic treated. Authors are free to use the material in other publications. However, if extensive use is made elsewhere, the publisher should be informed. Authors receive jointly 50 complimentary copies of their book. They are entitled to purchase further copies of their book at a reduced rate. As a rule no reprints of individual contributions can be supplied. No royalty is paid on Lecture Notes in Physics volumes. Commitment to publish is made by letter of interest rather than by signing a formal contract. Springer-Verlag secures the copyright for each volume.

The Production Process

The books are hardbound, and quality paper appropriate to the needs of the author(s) is used. Publication time is about ten weeks. More than twenty years of experience guarantee authors the best possible service. To reach the goal of rapid publication at a low price the technique of photographic reproduction from a camera-ready manuscript was chosen. This process shifts the main responsibility for the technical quality considerably from the publisher to the author. We therefore urge all authors to observe very carefully our guidelines for the preparation of camera-ready manuscripts, which we will supply on request. This applies especially to the quality of figures and halftones submitted for publication. Figures should be submitted as originals or glossy prints, as very often Xerox copies are not suitable for reproduction. In addition, it might be useful to look at some of the volumes already published or, especially if some atypical text is planned, to write to the Physics Editorial Department of Springer-Verlag direct. This avoids mistakes and time-consuming correspondence during the production period.

As a special service, we offer free of charge LaTeX and TeX macro packages to format the text according to Springer-Verlag's quality requirements. We strongly recommend authors to make use of this offer, as the result will be a book of considerably improved technical quality. The typescript will be reduced in size (75% of the original). Therefore, for example, any writing within figures should not be smaller than 2.5 mm.

Manuscripts not meeting the technical standard of the series will have to be returned for improvement.

For further information please contact Springer-Verlag, Physics Editorial Department II, Tiergartenstrasse 17, W-6900 Heidelberg, FRG.

Martin Schoen

Computer Simulation of Condensed Phases in Complex Geometries

Springer-Verlag Berlin Heidelberg GmbH

Author

Martin Schoen
Naturwissenschaftliche Fakultät, Universität Witten/Herdecke
Stockumer Strasse 10, W-5810 Witten, FRG
(new area code from 1.7.1993: 58453 Witten)

ISBN 978-3-662-13924-0   ISBN 978-3-540-47590-3 (eBook)
DOI 10.1007/978-3-540-47590-3

© Springer-Verlag Berlin Heidelberg 1993
Originally published by Springer-Verlag Berlin Heidelberg New York in 1993
Softcover reprint of the hardcover 1st edition 1993

Typesetting: Camera ready by author
58/3140-543210 - Printed on acid-free paper

Widmung *

Herbst und die dämmernden Sonnen im Nebel
Und nachts am Himmel ein Feuerbild.
Es stürzt und weht. Du mußt es bewahren.
Am Hohlweg wechselt schneller das Wild.
Und wie ein Hall aus fernen Jahren
dröhnt über Wälder weit ein Schuß.
Es schweifen wieder die Unsichtbaren
Und Laub und Blätter treibt der Fluß.

Der Jäger schleppt nun heim die Beute,
Das kiefernästig starrende Geweih.
Der Sinnende sucht andre Spur.
Er geht am Hohlweg still vorbei,
Wo goldner Rauch vom Baume fuhr.
Und Stunden wehn vom Herbstwind weise,
Gedanken wie der Vögel Reise,
Und manches Wort wird Brot und Salz.
Er ahnt, was noch die Nacht verschweigt,
Wenn in der großen Drift des Alls
Des Winters Sternbild langsam steigt.

This manuscript is dedicated
to
my parents
in appreciation
of
their endless love, care and support
and to
my friend Martin Tackenberg
whose loyal friendship
helped me through difficult times.

---

* P. Huchel, Ausgewählte Gedichte, Suhrkamp (Frankfurt, 1982)

Preface

Molecularly small confined phases play an important role in many scientific and engineering disciplines. For instance, the confining membrane of a living cell is known to affect the structure and transport of cellular water, which mediates the cell's metabolism and other biochemical processes. Transport of hazardous waste through the soil is strongly influenced by the adsorption of bulk phase molecules on the confining mineral surfaces. Finally, molecularly thin confined fluid films play a prominent part in lubrication. These examples illustrate the broad range of natural and commercial processes to which the present subject pertains.

Much experimental effort has been devoted to molecularly small confined phases, revealing the intriguing nature of such systems. Several sections of this book are therefore devoted to descriptions of experimental techniques.

To date even the most refined experiments do not yield direct information about structure and processes on the molecular scale. Computer simulations, on the other hand, do give such information and therefore complement real laboratory experiments. Several sections of this book discuss the link between experiments and the corresponding simulations.

Molecularly small confined phases are fascinating from a strictly theoretical viewpoint. Because of their confinement to spaces of molecular dimensions these systems are both inhomogeneous and anisotropic. The book discusses several examples to highlight the dramatic influence of inhomogeneity and anisotropy. They include the layering phenomenon, second order solid–liquid phase transitions and anisotropic classical (i.e. Markovian) as well as non-classical self-diffusion. From a more abstract perspective inhomogeneity and anisotropy may be viewed as a result of an external field superimposed on the condensed phase. The work presented should be relevant to a wide range of systems, regardless of the precise physical nature of the external field.

I intend this book to provide a timely and useful source of references to related work and have thus cited some 230 articles. I also hope that the book inspires further research by active workers and that it motivates others to become involved in the field.

In closing I would like to express my gratitude to those people who played central roles at various stages of my work. I am especially grateful to my teacher PD Dr. Claus Hoheisel (Ruhr-Universität Bochum), who introduced me to statistical physics and helped me to progress. I am indebted to Profs. John H. Cushman and Dennis J. Diestler (Purdue University) for our continuing collaboration and for the many stimulating discussions we have enjoyed over the years. Their constant support and personal friendship have contributed substantially to this work. I am particularly thankful for the many critical suggestions and comments that have improved this book. Last but not least, I thank Prof. Dr. Siegfried Hess (Technische Universität Berlin) for his immediate and generous support at a critical stage of this work.

In addition, I thank Purdue University Computing Center (PUCC) and the Scientific Council of Höchstleistungsrechenzentrum (HLRZ) at the Forschungszentrum Jülich for the generous allocation of computer time on the CYBER 205, ETA 10-P (PUCC) and the CRAY Y-MP/832 (HLRZ).

The book is a slightly abridged version of my *Habilitationsschrift*, which I submitted to the physics faculty of the Technische Universität Berlin in October 1992.

Witten, January 1993   *Martin Schoen*

Contents

1 Introduction

Phenomenological (i.e. "classical") thermodynamics represents one of the great achievements of nineteenth century physical science [1]. It is a complete and self-contained theory of thermal processes allowing one to predict the exact outcome of experiments. From just four elementary hypotheses (known as "The Laws of Thermodynamics") conclusions about thermal processes are drawn from rigorous mathematical operations [2].

Although mathematical rigour is one of the beauties of thermodynamics it also bears a certain disadvantage. Within the framework of the theory it can only be decided if a thermal process is possible; no statements are made about whether such a process will actually happen, i.e. the theory does not comment on mechanisms driving these processes. In other words, classical thermodynamics lacks a molecular basis.

This is not surprising because nineteenth century physical science did not immediately recognize atoms or molecules as real objects [3]. While "the atom" was a well established but purely philosophical entity around 1800 (to which consequently rather bizarre properties were ascribed [4]), it did not have precise meaning. Meaning was given to it in experiments reviewed by Lord Kelvin [4] in which atoms could be counted and weighed, so that the molecular structure of matter was soon to become an accepted fact among scientists. From there it was only a small step to realize the relation between heat (or temperature) and kinetic energy of individual particles [5], thereby putting thermodynamics conceptually on a firm molecular (i.e. microscopic) basis. In essence this approach enables one to compute thermodynamic functions like pressure, internal energy etc. from motions of individual particles.

Although the microscopic approach clearly enhances the level of description, it also poses a serious problem since any macroscopic piece of matter is composed of an astronomically large number of particles. Thus, following the recipe sketched above one would have to pursue the trajectory of each individual particle which is a hopeless venture even for very short times.

However, at the turn of the nineteenth century Maxwell [6] and Boltzmann [7] realized that with the introduction of statistical concepts a molecular approach might be preserved. Instead of studying the behavior of all particles individually, it suffices to employ a much smaller representative subgroup of them. The relevant dynamical information is then concealed in a function, say $f^{(N)}(\mathbf{q}^N, \mathbf{p}^N; t)$ which represents the probability (density) at time t to find the N particles of the subgroup at certain positions in space $\mathbf{q}^N$ and travelling with certain momenta $\mathbf{p}^N$. As time passes $f^{(N)}(\mathbf{q}^N, \mathbf{p}^N; t)$ changes according to some equation of "motion" and macroscopic properties (e.g. energy) can be expressed as time averages of their corresponding microscopic analogues (e.g. the system hamiltonian) where $f^{(N)}(\mathbf{q}^N, \mathbf{p}^N; t)$ provides proper weighting for the averaging process [8].

While this approach is general, Boltzmann restricted his theory originally to very dilute gases where the representative subgroup may be taken to consist of just two particles [9,10]. Temporal evolution then comes about via isolated binary collisions. Based on these ideas Boltzmann derived his famous kinetic equation which describes the change of $f^{(N)}(\mathbf{q}^N, \mathbf{p}^N; t)$ with time [9,10].

Historically the introduction of statistical concepts may be viewed as a reflection of laziness of the human mind. In Maxwell's and Boltzmann's days it was out of any question that *in principle one could* compute particle trajectories in a macroscopic system for arbitrarily long times and at any level of precision due to the classical perception of microscopic objects. Statistical concepts are then nothing but a matter of computational convenience. It was left to quantum mechanics in this century to prove that it is *in principle impossible* to determine the dynamics of even a single particle with arbitrary precision due to the Uncertainty Principle (see p. 40 in [1]). In recent years chaos theory could also demonstrate that in a deterministic (with regard to the equation of motion) classical system complete predictability of the system's future may be prohibited [11]. So from a modern perspective statistical physics takes cognizance of the *in principle incomplete* knowledge about microscopic dynamics as an immanent law of nature.

However, even without this more sophisticated modern view Boltzmann could close the gap between thermodynamics and a statistical description of matter by introducing a function H in terms of $f^{(N)}(\mathbf{q}^N, \mathbf{p}^N; t)$ which can be shown to remain constant or to decrease with time. It is, therefore, plausible to associate with H a thermodynamic state function, the entropy, so that the approach to equilibrium in a thermodynamic system can now be expressed in terms of probabilities of microscopic states $\{\mathbf{q}^N, \mathbf{p}^N, t\}$ [12]. From a microscopic perspective the increase of entropy in a dilute gas at non-equilibrium is then caused by a succession of binary collisions.

However, Boltzmann's approach was not immediately accepted by the scientific community [13,14] due to certain flaws in his argumentation (see also Chap. 5 in [3]). Nevertheless, the more general idea of a statistical description of matter outlasted the setbacks it suffered in its early days. Finally, in this century physicists became more used to a probabilistic description of nature due to the development and immediate success of quantum mechanics. Quantum mechanics eventually provided a sound basis for statistical physics as a whole [15].

Although the quantum-mechanically motivated revival reestablished statistical physics as a potentially powerful theory, it could only be applied with moderate success to more condensed states of matter for the first, say fifty years of this century. Perhaps, one of the most significant successes of statistical physics during that period is Debye's theory of heat capacity of solids [16]. Over a large range of temperatures down to the immediate vicinity of absolute zero this theory describes the temperature dependence of the heat capacity essentially in terms of the frequency spectrum of a continuous medium. Another important success was achieved by Onsager who solved analytically the two-dimensional Ising model of a magnet [17].

The application of statistical physics to dense fluids has been far less successful compared with both moderately dense gases and solids. The problem with dense fluids arises because of their typical density range many body effects become increasingly important. Fluids are also sufficiently disordered so that simplifying symmetry arguments (as for "lattice-based" systems) cannot be invoked (see [18]). Additional assumptions are indispensable to develop solvable analytical theories of dense fluids. Some of these assumptions are rather severe and, even worse, their consequences could be tested in the past to a very limited extent by comparison with only partially appropriate experimental data. An example is the so-called Kirkwood superposition approximation in the theory of fluid structure which expresses spatial correlations among a triplet of atoms as a product of three pair correlation functions each one involving a combination of two particles of the triplet [19]. It took about 30 years before the super-position approximation could finally be tested rigorously by means of computer simulations [20,21] in which the evolution of a microscopic model system is studied numerically.

With the advent of large-scale computers in the 1950's and sixties a new era began in statistical physics which is perhaps best termed "computational statistical physics". Computer simulations enabled scientists for the first time to treat complex many-particle systems on essentially a first principles basis so that assumptions involved in many already existing statistical-physical theories could rigorously be tested. However, computers were soon to become research tools in their own rights by which previously unknown and unexpected phenomena could be unveiled. The three, perhaps, earliest such discoveries concern the applicability of the concept of Brownian motion to self-diffusion on a microscopic time scale [22], the occurrence of hydrodynamic vortices in a microscopic system (see Fig. 1 in [23]) and the existence of liquid-solid phase transitions in hard sphere fluids [24].

Due to these and other remarkable successes computational methods are nowadays perceived as a third and independent branch in condensed matter physics, standing somewhere in-between the two classical ones, namely theory and experiment (see Fig. 1.2 in [25]). The still continuing rapid evolution of computational methods will depend largely on future advances in computer science and technology. However, since these two themselves are still rapidly developing fields it can be expected that computational statistical physics will continue to flourish and provide valuable insights into properties of condensed matter. In fact, the enhancement of computational speed over the past 10–15 years due to the advent of vector and parallel computers rendered possible investigations of increasingly complex phenomena and systems. In turn, results of these studies lead to new theoretical concepts in condensed matter physics.

It may therefore be appropriate to describe the current status of statistical physics as a "Golden Age" in Truesdell's sense [26]. According to his philosophy there is a period of time (i.e. the "Golden

Age") during which all the significant developments in a scientific discipline are made; after that period it only remains to study the works of the "founding fathers" repeatedly to fill the marginal gaps they might have overlooked accidentally. Applying Truesdell's Renaissance humanist point of view [3] to statistical physics as it appears today, one is tempted to conclude that its "Golden Age" may still be lingering.

The following article tries to demonstrate this by discussing behavior and properties of vicinal phases. A vicinal phase consists of an assembly of particles (atoms or molecules) confined by solid surfaces (i.e. walls). The walls render vicinal phases inhomogeneous, i.e. density becomes a function of spatial position. Depending on their own inherent symmetry the walls may also cause vicinal phases to be anisotropic in one or more spatial directions.

From an abstract perspective the walls may be represented as an external, symmetry-breaking field. Following this notion vicinal phases are nothing but a special realization of a more general class of systems in which matter is exposed to some external (electric, magnetic, gravitational etc.) field. It is therefore not unreasonable to expect the results to be presented below to pertain qualitatively to a wide range of different systems.

Depending on a variety of parameters such as thermodynamic conditions or geometric restrictions, vicinal phases may exhibit gas-, liquid- or solidlike characteristics. However, confinement to spaces of molecular dimensions by the walls affects properties of vicinal phases in a way that the above states of matter are distinguished from those in corresponding bulk phases in a unique and sometimes dramatic way [27].

Hence the behavior of vicinal phases is of vital importance for many natural and commercial processes. In biological systems, vicinal water [28] is speculated to influence growth, metabolism and intracellular organization [29]. It has been implicated in the folding of DNA molecules [30]. In the natural environment, vicinal fluids play a crucial role in adsorption and transport in clay soils [31] and microporous media in general [32] consequently determining stability of earthen structures and dispersion of environmental pollutants. A number of processes in engineering, such as lubrication [33], catalysis [34] and drilling [35] involve vicinal fluids.

While this short and necessarily incomplete list highlights the importance of vicinal phases from a practical point of view, it is the purpose of this article to demonstrate their even more fascinating characteristics from a theoretical perspective. In this respect computer simulation methods such as molecular dynamics (MD) and Monte Carlo (MC) calculations play a prominent role because they yield experimentally inaccessible microscopic information about vicinal phases [36].

This is exemplified in Chap. 5 which is predominatly concerned with self-diffusion in vicinal fluids. MD permits investigations of diffusion as a function of position of the diffusing atom between the walls which elucidates many details about the transport of matter in confined phases. In MD one may also study diffusion in different spatial directions which permits an explanation of inhomogeneity and anisotropy as temporal phenomena. In addition, from MD results quantitative conclusions can be drawn about the validity of hydrodynamic concepts on which experimental pictures of diffusion are generally based [37]. The computer simulation results will also be shown to give evidence of anomalous, i.e. non-Fickian diffusion in a regime prior to solidification of vicinal fluids.

Solidification in general can occur as a first or as a second order phase transition in vicinal phases. The order of the transition depends on the "driving force" as will be shown in Chap. 4. However, a second order liquid-solid phase transition is a unique characteristic of thin vicinal phases and must not occur in the bulk on account of the symmetry changes involved [38].

The general conditions under which solid- and liquidlike vicinal phases may exist are discussed in Chap. 3. This chapter is also devoted to a discussion of the layering phenomenon which is another unique feature of vicinal phases. Layering was first suggested indirectly by experiments but only later verified directly in computer simulations.

However, it seems to be both pedagogically expedient and organizationally convenient to begin the discussion in Chap. 2. with a brief introduction to computer simulation methods relevant to vicinal phase studies.

2 Computer Simulation Methods

2.1 Preliminary Remarks

Computer simulations have the advantage that they provide essentially exact results for a chosen model system because they involve only a very limited and easy to control number of additional approximations. Probably the most significant one concerns the choice of an interaction model when experimental systems are to be represented (see Chap. 3).

The present section is devoted to a very brief introduction to computer simulation methods employed in vicinal phase studies. The discussion begins in the subsequent section with the MD method [39]; a description of various MC methods is deferred to Sect. 2.3. The advantage of MD is simultaneous accessibility of time-dependent *and* equilibrium properties. The application to time-dependent properties will be discussed in detail in Chap. 5. The MC method, on the other hand, has the advantage that it may easily be applied to various statistical-physical ensembles and to quantum-mechanical problems [40]. However, the present article is exclusively concerned with systems which may be treated classically (i. e. "classical systems").

2.2 Molecular Dynamics

2.2.1 The Equation of Motion in a Classical Many-Particle System

Trajectories of individual classical particles are governed by Newton's equation of motion

$$-\frac{\partial \Phi\left(\mathbf{r}^N(t)\right)}{\partial \mathbf{r}_i(t)} = \mathbf{F}_i\left(\mathbf{r}(t)\right) = m_i \ddot{\mathbf{r}}_i(t) \tag{2.1}$$

Eq. (2.1) assumes a conservative system where m_i is the mass, $\ddot{\mathbf{r}}_i$ is the acceleration of particle i, Φ is the potential energy and $\mathbf{r}^N$ is an abridgement for the set of (Cartesian) coordinates $\{\mathbf{r}_1, \mathbf{r}_2, \ldots, \mathbf{r}_N\}$. $\mathbf{F}_i$ denotes the total force exerted on particle i by the remaining $N-1$ particles and the walls.

Eq. (2.1) is a system of coupled second order differential equations. Coupling arises through the total force on particle i, $\mathbf{F}_i$, which depends on the entire set $\mathbf{r}^N$ at any instant. Clearly, even for small N analytical solutions of eq. (2.1) can hardly be obtained but numerical solutions are accessible by procedures to be described in the following sections. Numerical techniques to solve eq. (2.1) are referred to as equilibrium MD (i. e. EMD) and static properties can be computed as temporal averages $\langle O \rangle_t$ by averaging a suitable microscopic representation $O(\mathbf{r}^N(t), \mathbf{p}^N(t))$ of the quantity O over phase space trajectories generated via eq. (2.1).

An important feature of eq. (2.1) is conservation of energy. Phase space trajectories generated via eq. (2.1) are then restricted to hyperplanes of constant energy E, i. e. microstates $\{\mathbf{p}^N(t), \mathbf{r}^N(t)\}$ are representative of the microcanonical ensemble. This is, however, not quite correct because eq. (2.1) also conserves linear momentum in the absence of external fields. Rigorously, microstates from eq. (2.1) represent only a subensemble of the microcanonical ensemble but this distinction is minor and of no practical concern [25]. As pointed out in Chap. 1, the walls confining vicinal phases are nothing but a representation of such external fields. Thus, linear momentum of vicinal phase particles cannot be conserved whereas the total energy must still be constant as long as there are no explicitly time-dependent force contributions.

One may, however, have such non-conservative, explicitly time-dependent force contributions so that the system does not conserve energy. Numerical methods solving the equation of motion subject to non-conservative forces are usually referred to as non-equilibrium MD (i.e. NEMD) [41]. Applications to vicinal phases are discussed in Sect. 4.3.3 and 5.4.

2.2.2 Numerical Aspects

Continuous potentials. For a continuously differentiable intermolecular potential (i.e. "soft" potential) the physical trajectory of a particle in a conservative system is approximated in EMD by applying a finite difference scheme to eq. (2.1). For instance, approximating $\ddot{\mathbf{r}}_i(t)$ by

$$\ddot{\mathbf{r}}_i(t) = \frac{\mathbf{r}_i(t+\delta t) - 2\mathbf{r}_i(t) + \mathbf{r}_i(t-\delta t)}{(\delta t)^2} \tag{2.2}$$

permits one to rewrite eq. (2.1) as

$$\mathbf{r}_i(t+\delta t) = 2\mathbf{r}_i(t) - \mathbf{r}_i(t-\delta t) + \frac{\mathbf{F}\left(\mathbf{r}_i^N(t)\right)}{m_i}(\delta t)^2 \tag{2.3}$$

Eq. (2.3) is the so-called "Störmer-Verlet algorithm". The Störmer-Verlet algorithm is employed in Chap. 5 for vicinal phase models to be introduced in Sect. 3.3.1 and 3.3.2. There are, of course, numerous other numerical schemes to solve eq. (2.1) and the interested reader is referred to [25,42].

For practical applications it is noteworthy that the computation of the set of forces $\mathbf{F}^N(\mathbf{r}^N(t))$ is the most time-consuming step in MD. It requires typically 90-95% of the entire CPU time so that efficient coding of the force calculation is indispensable. In the past various schemes have been proposed for an efficient computation of the forces [43-50]. These schemes aim at a high degree of vectorizability of the force calculation but differ conceptually and with respect to numerical efficiency [44,47]. A particularly simple technique has been proposed by Schoen [44]. Although this method is not the most efficient one [47], its application to vicinal phases is straightforward and self-explanatory from the FORTRAN source code presented for homogeneous bulk systems in [44].

In addition to the set of forces the iterative solution of eq. (2.3) requires two subsequent configurations in time, namely $\mathbf{r}^N(t)$ and $\mathbf{r}^N(t - \delta t)$. A small conceptual problem arises for the very first time step where a second configuration $\mathbf{r}^N(t - \delta t)$ to initiate the iteration process is not readily available. However, from

$$\mathbf{r}_i(t - \delta t) = \mathbf{r}_i(t) - \mathbf{v}_i \delta t \tag{2.4}$$

this second configuration may be obtained provided the set of velocities $\mathbf{v}^N$ is known. Components of $\mathbf{v}_i$ may be chosen at random from a Maxwell-Boltzmann distribution at the desired temperature. Except for this initiation of the iteration process velocities do not enter the algorithm but may be obtained from

$$\mathbf{v}_i(t) = \frac{\mathbf{r}_i(t + \delta t) - \mathbf{r}_i(t - \delta t)}{2 \delta t} \tag{2.5}$$

They are important with regard to temperature T which is computed via the equipartition theorem. Velocities may also be required with respect to self-diffusion as will be shown in Chap. 5.

Finally, accuracy of the Störmer-Verlet algorithm depends critically on the characteristic length of the "finite difference", i.e. on the size of the time step δt. A too large δt results in dissatisfactory energy conservation. Energy conservation should therefore be monitored in any EMD calculation to avoid insufficiently accurate approximations of the "true" trajectories which may lead to nonsensical physical results. In most applications $\delta t \approx 10^{-14}$s is sufficient to conserve energy to better than 0.1%.

Partially discontinuous potentials. While the method outlined so far cannot be applied to purely discontinuous potentials (see Sect. 3.6.1 in [25]) on account of the discontinuity in $\mathbf{F}_i$, it can be modified to handle forces of very different ranges. This situation arises when "soft" vicinal particles are confined by "hard" walls (see Sect. 3.3.3, 3.3.4. and 5.2.3.) [51].

As long as particles do not collide with a hard wall, eq. (2.3) is solved iteratively. At every time step it is then checked whether any particles attempted to cross a wall during the interval $[t, t+\delta t]$. Suppose a subgroup of n particles attempted to cross a wall, eq. (2.3) describes the trajectory of particle i(n) up to time $t+\delta t_i < t+\delta t$ (i(n) indicates that particle i belongs to the subgroup of particles attempting to cross a wall). $\{\delta t_i\}$ may be computed from the known position of the walls provided the velocity of crossing is known. This is approximated from the virtual position of particles behind the wall at time $t+\delta t$ and the position at time $t-\delta t$ from eq. (2.5). If δt is small enough $(5 \cdot 10^{-15} s)$ the deviation between the velocity at time t from eq. (2.5) and the correct velocity at time $t+\delta t_i$ can be assumed to be small enough so that the numerical trajectory does not deviate too much from the physically correct one. This is again established from the criterion of energy conservation which must hold for discontinuous potentials, too. In practice, energy is conserved here to 0.1%.

Upon collision with the wall each particle is scattered back into the vicinal phase. The new velocity after interaction with the wall, $\mathbf{v}_i'(t+\delta t_i)$, is determined according to the principle of energy conservation and the physical nature of the walls (see Sect. 3.3.3 and 3.3.4.). Since δt is fixed each particle may move for a remaining time of

$$\delta t_i' = \delta t - \delta t_i \tag{2.6}$$

after having hit the wall. Its new position is then obtained from

$$\mathbf{r}_i'(t+\delta t) = \mathbf{r}_i(t+\delta t_i) + \mathbf{v}_i'(t+\delta t_i)\delta t_i' \tag{2.7}$$

To restart the Störmer-Verlet algorithm one also needs the virtual position behind the wall $\tilde{\mathbf{r}}_i'(t)$. It is obtained by analogy with eq. (2.5) as

$$\tilde{\mathbf{r}}_i'(t) = \mathbf{r}_i'(t+\delta t) - \mathbf{v}_i'(t+\delta t_i)\delta t_i \tag{2.8}$$

In the next iteration $\tilde{\mathbf{r}}_i'(t)$ and $\mathbf{r}_i'(t+\delta t)$ replace $\mathbf{r}_i(t-\delta t)$ respectively $\mathbf{r}_i(t)$ in eq. (2.3). For further details see [51].

2.3 Monte Carlo Methods

The microcanonical ensemble mentioned in the previous section is not of prime interest in vicinal phase studies. This is due to key experiments one wishes to connect with. In these experiments thermodynamic states are not characterized by constant N, V and E but some other set of state parameters (see Sect. 3.2, 4.3).

However, it may be difficult to solve the equation of motion in more relevant statistical-physical ensembles. This is particularly true for the grand canonical ensemble in which the thermodynamic state is characterized by constant chemical potential µ whereas N may vary. A change in N, on the other hand, may cause the forces to fluctuate too strongly so that any finite difference approximation to eq. (2.1) is very likely to break down.

However, restricting the discussion to static properties one may abandon a numerical solution of eq. (2.1) and compute these properties as ensemble averages $\langle O \rangle_e$ instead of temporal averages $\langle O \rangle_t$ as in Sect. 2.2 if the ergodic hypothesis [8,41] is valid. In principle, $\langle O \rangle_e$ can be obtained by performing a random walk in configuration space [52]. Based upon the theory of Markov processes [53] this may be done in MC calculations provided the probability density in configuration space is known [8,54,55].

2.3.1 The Modified Metropolis Algorithm in the Grand Canonical Ensemble

In conjunction with experiments described in Sect. 3.2 the grand canonical ensemble is of prime importance with respect to equilibrium properties of vicinal phases.. The grand canonical ensemble represents an open system in which a thermodynamic state is specified by chemical potential µ, V and T. So in addition to coupling to an infinitely large external heat reservoir (closed system, canonical ensemble), the grand canonical ensemble is coupled to a similar reservoir of matter. Using, for instance, Schrödinger's method of lagrangian multipliers [55] the probability density (classical limit!) in the grand canonical ensemble may be written as (for details, see also Sect. 3-1 in [8])

$$f_0^{(N)}(\mathbf{r}^N) = \left(Q_{\mu VT}\right)^{-1} e^{(BN - \ln N!)} e^{-\beta \Phi(\mathbf{r}^N)} \qquad (2.9)$$

where

$$B = \beta\mu - \ln\left(\Lambda^3/V\right) \qquad (2.10)$$

in the notation suggested by Adams [56]. $\Lambda = \left(h^2\beta/2\pi m\right)^{1/2}$ is the thermal de-Broglie-wavelength with h being Planck's constant and $Q_{\mu VT}$ is the grand canonical ensemble partition function. Eq. (2.9) involves an integration over momentum subspace which can be carried out analytically for a conservative system in which momenta are distributed according to a Maxwell-Boltzmann distribution. The Maxwell-Boltzmann distribution is preserved in vicinal phases despite the presence of the external field, i. e. the walls [57].

To generate a sequence of configurations complying with $f_0^{(N)}$ in eq. (2.9) a random walk in configuration space is carried out. The random walk is conducted as a Markov process because $Q_{\mu VT}$ is unknown during the MC calculation For a Markov process it may be shown [58,59] that only the *relative* probability of occurrence of two states m and n is required.

In the grand canonical ensemble a numerical representation of a Markov chain of particle configurations may be generated by a modified version of the classical Metropolis algorithm [60]. The modified Metropolis algorithm is carried out as a pair of consecutive steps. During the first step N (i. e. $N_m = N_n = N$) is fixed and the system evolves in configuration space by particle displacements. Since N stays fixed the first step is identical with the classical Metropolis algorithm in which stochastic diffusional motion is the only dynamical event. The relative probability of occurrence is then given by

$$\frac{f_0^{(N)}(\mathbf{r}_n^N)}{f_0^{(N)}(\mathbf{r}_m^N)} = \frac{\exp\left[-\beta\Phi(\mathbf{r}_n^N)\right]}{\exp\left[-\beta\Phi(\mathbf{r}_m^N)\right]} = \exp\left[-\beta\,\Delta\Phi_{mn}\right] \qquad (2.11)$$

The classical Metropolis algorithm may be realized numerically as follows. Starting from an arbitrary arrangement one begins by picking one of the N particles at random or sequentially and computes its potential energy Φ_m. The particle is then displaced according to

$$\mathbf{r}_{i,n} = \mathbf{r}_{i,m} + \delta(1-2\boldsymbol{\xi}) \qquad (2.12)$$

where $\mathbf{r}_{i,m}$ and $\mathbf{r}_{i,n}$ are the particle's old respectively new positions, δ is the side length of a small cube centered on $\mathbf{r}_{i,m}$, $\mathbf{1}$ denotes the vector $(1,1,1)$ and $\boldsymbol{\xi}$ is a random vector whose three components are pseudo-random numbers from a uniform distribution on the interval $[0,1]$. δ is adjusted during the MC run according to the criterion of 40-60% acceptance of attempted displacements (see p. 122 of [25]).

At the new position the particle's potential energy is recalculated (Φ_n). In the Metropolis algorithm displacements are accepted based upon the criterion (see eq. (2.11))

$$\Pi_1 = \begin{cases} 1 & ; \ \Phi_m \geq \Phi_n \\ \exp\left[-\beta\,\Delta\Phi_{mn}\right] & ; \ \Phi_m < \Phi_n \end{cases} \qquad (2.13)$$

From eq. (2.13) it follows that displacements are accepted with a probability of $\Pi_1 = 1$ if they lower Φ; if they enhance Φ, they are accepted only with a lower probability equal to the Boltzmann factor in the second line of the r.h.s. of eq. (2.13). The latter is realized in practice by drawing a pseudo-random number ξ from the uniform distribution on $[0,1]$ and comparing it with Π_1. If $\Pi_1 > \xi$, the displacement is still accepted; if $\Pi_1 \leq \xi$, the displacement is finally rejected. The old configuration with the selected particle at $\mathbf{r}_{i,m}$ is reestablished and the procedure starts all over again.

During the second step of the Markov chain generation N may vary. Obviously, a change in N can result from attempted additions or removals of particles. Assuming $\Delta N_{mn} = \pm 1$, it follows from eq. (2.9) that

$$\frac{f_0^{(N+1)}(\mathbf{r}_n^{N+1})}{f_0^{(N)}(\mathbf{r}_m^N)} = \exp(r_+) \tag{2.14a}$$

for addition, while for removal

$$\frac{f_0^{(N-1)}(\mathbf{r}_n^{N-1})}{f_0^{(N)}(\mathbf{r}_m^N)} = \exp(r_-) \tag{2.14b}$$

with

$$r_+ = B - \ln N_n - \beta \Delta\Phi_{mn} \tag{2.15a}$$

and

$$r_- = -B + \ln N_m + \beta \Delta\Phi_{mn} \tag{2.15b}$$

$\Delta\Phi_{mn}$ in eq. (2.15a) is the change in potential energy due to addition of a particle at a randomly chosen point in space. Since only one particle is added, $\Delta\Phi_{mn}$ is identified with that particle's potential energy. Similarly, $\Delta\Phi_{mn}$ in eq. (2.15b) denotes the potential energy of a randomly selected particle earmarked for removal. By analogy with eq. (2.13) addition or removal of a particle is accepted according to the modified Metropolis criterion

$$
\Pi_2 = \begin{cases} 1 & ; \, r_{+,-} > 0 \\ \exp\left(r_{+,-}\right) & ; \, r_{+,-} \leq 0 \end{cases} \tag{2.16}
$$

If r_+ (r_-) is less than or equal to zero addition (removal) of a particle is accepted if $\Pi_2 > \xi$, where ξ denotes a pseudo-random number from a uniform distribution on [0,1] as before. In the grand canonical (i. e. addition (removal)) substep it is decided with equal probability whether an attempted addition or removal of a particle is to be realized.

From a computational point of view the grand canonical substep differs from the diffusional one because for the former there is no way to adjust the acceptance rate as is for the latter by varying δ. For the grand canonical substep the acceptance rate is solely determined by the system's density and can be as small as 0.1% at very high densities. Thus, very long MC runs are required at high densities to guarantee for a limiting distribution of states proportional to eq. (2.9).

2.3.2 The Modified Metropolis Algorithm in the Isostress-Isostrain Ensemble

In certain experiments to be discussed in Sect. 4.3.1 solidlike vicinal phases are subjected to compressional and shear forces. Increasing the latter beyond a limiting threshold causes vicinal phases to undergo solid-liquid phase transitions. Under certain conditions these phase transitions are characterized by lack of concomitant drainage, i.e. $N = \text{const.}$. To investigate solid-liquid phase transitions of this nature in computer simulations (see Sect. 4.3) it is convenient to introduce an ensemble in which N particles are placed in a rectangular container at fixed T; compressional and shear forces can then be discussed in terms of movements of container walls. Quantitatively, in the limit of infinitesimally small compression or shear such movements can be described in terms of the displacement gradient tensor $\boldsymbol{u}$ Its elements $u_{\alpha\beta}$ specify the relative separation of elements of mass under a prescribed strain [61-63]. Spatial positions of a mass element in the deformed state, $\mathbf{r}$, and in the unstrained reference state, $\mathbf{r}^0$, respectively are related by

$$\mathbf{r} = \tilde{D}\,\mathbf{r}^0 \tag{2.17}$$

where $\tilde{D}$ is a matrix of constants if the displacement of the mass element, $\mathbf{v} = \mathbf{r} - \mathbf{r}^0$, is homogeneous. Consequently,

$$\mathbf{v} = (\tilde{D} - \mathbf{1})\mathbf{r}^0 \tag{2.18}$$

where $\mathbf{1}$ is a matrix whose α,β elements are $\delta_{\alpha\beta}$ (the Kronecker delta). As an example consider a rectangular prism of material constrained in the octant $(x>0,\ y>0,\ z>0)$ with its faces coincident with the planes $x=0$, $y=0$ and $z=0$. Suppose its (reference) volume is $V^0 = s_x^0 s_y^0 h^0$. Now imagine that the prism is stretched or compressed in the x-, y- and z-directions to a new volume $V=s_x s_y h$ and then sheared isochorically by moving the face lying in the plane $z=h$ a distance αl in the x-direction. As will be shown in Sect. 3.3.1 the term αl specifies relative alignment of the container (i. e. prism) walls; an explanation of the physical significance of α and l is also deferred to that section.

A mass element at $\mathbf{r}^0$ in the reference state is located at $\mathbf{r}$ in the strained state. Thus,

$$\begin{pmatrix} r_1 \\ r_2 \\ r_3 \end{pmatrix} = \begin{pmatrix} s_x/s_x^0 & 0 & \alpha l/h^0 \\ 0 & s_y/s_y^0 & 0 \\ 0 & 0 & h/h^0 \end{pmatrix} \tag{2.19}$$

$$= \tilde{D}\,\mathbf{r}^0$$

In this case $\mathbf{v}$ is given by

$$\begin{pmatrix} v_1 \\ v_2 \\ v_3 \end{pmatrix} = \begin{pmatrix} (s_x - s_x^0)r_1^0/s_x^0 + \alpha l r_3^0/h^0 \\ (s_y - s_y^0)r_2^0/s_y^0 \\ (h - h^0)r_3^0/h^0 \end{pmatrix} \tag{2.20}$$

Defining components of the displacement gradient tensor $\boldsymbol{u}$ as $u_{\alpha\beta} \equiv \partial v_\alpha / \partial r_\beta^0$ it follows from eq. (2.20) that

$$\boldsymbol{u} = \begin{pmatrix} \sigma_1 & 0 & \sigma_4 \\ 0 & \sigma_2 & 0 \\ 0 & 0 & \sigma_3 \end{pmatrix} \qquad (2.21)$$

where the strains are given by

$$\sigma_1 = (s_x - s_x^0) / s_x^0 \qquad (2.22a)$$

$$\sigma_2 = (s_y - s_y^0) / s_y^0 \qquad (2.22b)$$

$$\sigma_3 = (h - h^0) / h^0 \qquad (2.22c)$$

$$\sigma_4 = \alpha l / h^0 \qquad (2.22d)$$

Eqs. (2.22a) - (2.22c) describe compression of the container while eq. (2.22d) describes shear in x-direction applied to planes parallel to the z-direction of the laboratory coordinate frame. Notation for the stresses parallels that for the strains, i. e. $\tau_1 = T_{xx}$, $\tau_2 = T_{yy}$ and so forth. $T_{\alpha\beta}$ is the α-th (Cartesian) component of the force exerted on a unit area pointing in β-direction. According to eq. (2.21) $\boldsymbol{u}$ is a second rank tensor with nine components and volume V of the strained container is related to the unstrained volume $V^0 = s_x^0 s_y^0 h^0$ by

$$V \approx V^0 (1 + \operatorname{Tr} \boldsymbol{u}) \qquad (2.23)$$

under the proviso of infinitesimal strains.

In what follows a special case of this more general situation will be considered. It is assumed that the container is contracted or expanded only in z-direction subject to some constant stress τ_3 , s_x and s_y being held fixed and set to their reference values s_x^0 respectively s_y^0. In addition it will be assumed that $s_x = s_y = s$. Consequently, σ_3' is constant where σ_3' is a vector with three components given by eqs. (2.22a), (2.22b) and (2.22d). From the detailed thermodynamic analysis in [64] it is straightforward to write down an expression for the exact differential of the relevant thermodynamic potential, dG, in terms of N, T, τ_3 and σ'_3 which define the thermodynamic state of the vicinal phase in what shall henceforth be called the "isostress-isostrain ensemble".

It is shown in [64] that

$$\langle O \rangle_e = \frac{\int dh\, h^N \exp(\beta\tau_3 s^2 h) \int d\mathbf{r}'^N O(\mathbf{r}'^N;\boldsymbol{\sigma})\exp\left[-\beta\Phi(\mathbf{r}'^N;\boldsymbol{\sigma})\right]}{\int dh\, h^N \exp(\beta\tau_3 s^2 h) \int d\mathbf{r}'^N \exp\left[-\beta\Phi(\mathbf{r}'^N;\boldsymbol{\sigma})\right]} \qquad (2.24a)$$

after an integration over momentum subspace. By analogy with eq. (2.9) the isostress-isostrain ensemble probability density in configuration space is then given by

$$f_0^{(N)'}(\mathbf{r}^N;\boldsymbol{\sigma}) = \left(Q_{N\tau_3 T}\right)^{-1} h^N e^{\beta\tau_3 s^2 h} e^{-\beta\Phi(\mathbf{r}'^N;\boldsymbol{\sigma})} \qquad (2.24b)$$

where the unknown partition function $Q_{N\tau_3 T}$ is given by the denominator in eq. (2.24a). In both expressions $\boldsymbol{\sigma}$ is a vector whose four components are given by eqs. (2.22a) – (2.22d). Note, that eq. (2.24a) involves a convenient coordinate transformation

$$\mathbf{r} \longrightarrow \mathbf{r}' = \begin{pmatrix} x' \\ y' \\ z' \end{pmatrix} = \begin{pmatrix} x/s \\ y/s \\ z/h \end{pmatrix} \qquad (2.25)$$

which causes the factor h^N in eqs. (2.24a) and (2.24b). The transformation is computationally convenient because then the second integration in numerator and denominator on the r. h. s. of eq. (2.24a) extends only over the constant unit cube volume. An additional factor s^{2N} resulting from the coordinate transformation (eq. (2.25)) cancels between numerator and denominator in eq. (2.24a) due to the constancy of s.

As before in the grand canonical ensemble, the generation of a Markov chain may proceed in two steps. One of these consists of changing h by a small amount via

$$h_n = h_m + \delta_h (1 - 2\xi) \tag{2.26}$$

analoguous to eq. (2.12). The new configuration is obtained by scaling the z-component of vicinal particle and wall coordinates by the ratio h_n/h_m. As before the acceptance rate is adjusted here to 40-60% by adjusting δ_h during the MC run. Typical values of δ_h^* (see Table 3.1 on p. 39 for reduced units) resulting from this criterion for vicinal phases are $0.1 - 0.3$ [64]. The analogue of eqs. (2.11), (2.14a) or (2.14b) is then given by

$$\frac{f_0^{(N)'}(\mathbf{r}_n^{'N}, h_n)}{f_0^{(N)'}(\mathbf{r}_m^{'N}, h_m)} = \exp(r_h) \tag{2.27}$$

and it can easily be verified from eq. (2.24b) that

$$r_h = \beta\left(\tau_3 s^2 \Delta h - \Delta\Phi_{mn}\right) + N \ln\left(h_n/h_m\right) \tag{2.28a}$$

$$\Delta h = h_n - h_m \tag{2.28b}$$

$\Delta\Phi_{mn}$ is the change in potential energy due to contraction (expansion) in z-direction. The modified Metropolis algorithm accepts a change in h with probability

$$\Pi_2 = \begin{cases} 1 & ; \; r_h > 0 \\ \exp(r_h) & ; \; r_h \leq 0 \end{cases} \qquad (2.29)$$

where the second condition is fulfilled by comparing $\exp(r_h)$ with a uniformly distributed pseudo-random number as before (see Sect. 2.3.1).

As before in the grand canonical ensemble one might perform purely diffusive steps in addition to the contraction (expansion) step just described. During the diffusive step h is fixed and particle displacements are accepted with probability Π_1 (eq. (2.13)) which is recovered from eqs. (2.27) - (2.29) if $h_m = h_n$.

In practice, the ratio between isostress-isostrain and diffusive sub-steps is taken to be 1:N on account of the much larger consumption of CPU time of the former. This is due to the scaling of z-coordinates of all particles and the walls by h_n/h_m which affects all contributions to Φ simultaneously. Since only the z-components of all coordinates in state n differ by the factor h_n/h_m from those in state m, it is impossible to obtain Φ_n by a scaling argument from Φ_m so that a double calculation of the *total* potential energy is required during each contraction (expansion) step. Restricting the discussion to pairwise additive potentials, the number of arithmetic operations during contraction (expansion) is of the order of $O(N^2)$ while during the diffusive step this number is of the order of $O(N)$ because only coordinates of a single particle are affected. Therefore, isostress-isostrain ensemble MC is computationally more demanding than MC in the isothermal-isobaric ensemble which is often employed to study phase equilibria in homogeneous bulk phases [25,58]. In the latter $r_n'^N$ during contraction (expansion) is obtained by scaling all three (Cartesian) components of $r_m'^N$ uniformly by $(V_n/V_m)^{1/3}$ where the volumes V_m and V_n are related by an expression similar to eq. (2.26). For pairwise additive potentials Φ_n can then be obtained by appropriately scaling Φ_m at half the computational effort compared with isostress-isostrain ensemble MC.

3 Equilibrium Properties of Vicinal Phases

3.1 Preliminary Remarks

In the preceding chapter aspects of computer simulation methods relevant to vicinal phase studies are briefly discussed. In the present and the two following chapters computer simulations are now employed to study a special class of systems in condensed matter physics. These systems are distinguished by both inhomogeneity and anisotropy. Inhomogeneity becomes manifest structurally in terms of a spatially varying density. Anisotropy like inhomogeneity is a result of the presence of an external symmetry-breaking field.

While external fields regardless of their precise nature will always cause these effects, this article focusses exclusively on vicinal phases as a special realization of inhomogeneous and anisotropic systems. In a vicinal phase the external field is represented by the confining walls.

Modern technology permits experimental studies of vicinal phases on a truly microscopic scale. However, although these experiments reveal a lot of remarkable information about unique features of vicinal phases, their scope is still rather limited. For example, at present no experiment can provide any direct information about vicinal phase structure.

Computer simulation methods are ideally suited to complement experiments because structure of vicinal phases is directly accessible to them. In addition, computer simulations yield results for proper-ties which can also be measured. Despite the simple prototypical model systems employed computationally, these results are in excellent qualitative agreement with the much more complex systems investigated experimentally. An example will be discussed in Sect. 3.4.1. The close connection between simulation and experiment enables relevant theoretical work with reasonable computational effort.

3.2 Experimental Background

In general, structure of condensed phases can be determined experimentally by X-ray or neutron scattering techniques [65]. While these methods are nowadays routinely used to determine bulk phase structure [65] or even structures of adsorbates on solid substrates [66], vicinal phases are inaccessible to these techniques at present.

However, experimental data permit indirect conclusions about vicinal phase structure. The basic idea behind those indirect approaches is to relate structure to the net force exerted on the wall by the vicinal phase. Since one is dealing with rather weak forces very sensitive measuring and control devices are indispensable. The three most fruitful experiments in such force measurements employ the surface force apparatus (SFA) [67], total internal reflection microscopy (TIRM) [68] and the atomic force microscope (AFM) [69]. They are briefly summarized in the following sections.

3.2.1 The Surface Force Apparatus

Experimental setup. Probably the most prominent experiment to determine surface forces employs the SFA. A schematic representation of the experimental setup can be found in Fig. 10.7 of [70].

The central part of the SFA consists of two mica surfaces in crossed cylinder configuration. Mica can be prepared with atomic smoothness over molecularly large areas. The crossed cylinders are immersed in a reservoir of bulk fluid so that molecules can flow from the bulk in-between the surfaces until thermodynamic equilibrium is reached. By means of the Derjaguin approximation [71] the crossed cylinder configuration can be translated locally into a two-spheres arrangement. Since the radii of the cylinders are macroscopic (ca.

10^{-2} m), the "contact" area of the corresponding spheres is approximately flat on a microscopic length scale.

Attached to the cylinders are interchangeable force-measuring springs of known stiffness. Separation between the cylinders can be changed from microns down to molecular contact by means of a triple step device, the last and most sensitive one involving a piezo-electric crystal tube. This allows vertical positioning of the cylinders with an accuracy of 0.1 nm. The actual distance between the cylinder surfaces is measured optically using interference fringes. Positions and shapes of the (colored) fringes allow for a determination of surface separation to better than 0.1 nm. The amount of matter between the cylinders can be determined by measuring the refractive index of the vicinal phase.

Given the precision of the surface separation measurement the corresponding force measurement is straightforward. The piezoelectric crystal is compressed or expanded by a known amount. It is then determined optically how much the separation between the cylinder surfaces actually changed. From the difference between the two values the associated force changes can be inferred directly. This is achieved by multiplying the length difference by the known stiffness of the force measuring spring. Sensitivity of force measurements is typically of the order of $O(10^{-8} \text{N})$ [70]. With the SFA both repulsive and attractive forces can be measured and the interested reader is referred to Israelachvili's book [70] for further technical details and a comprehensive review of systems and problems studied so far by this technique.

An Illustration: Organic Vicinal Phases. Typical experimental results obtained with the SFA are shown in Fig. 3.1 for various organic fluids between mica surfaces [72]. The curves for cyclohexane, tetradecane and hexadecane reveal a lot of oscillatory structure as surface separation changes. As will be shown later in Sect. 3.4.2 oscillations in the force curves signify layering of vicinal phase particles parallel with the walls. It should, however, be noted again that there is no quantitative experimental link between force measurements and local vicinal phase structure although other independent experiments [73-76] support the notion of layering of vicinal phases, too.

A second noteworthy point about Fig. 3.1 concerns peak positions. While cyclohexane, tetradecane and hexadecane all exhibit oscillatory curves, molecular structure makes a difference as far as chain versus

26

planar geometries are considered; chain length apparently is irrelevant. Based upon the assumption of layering one may speculate that commensurability of vicinal phase molecular and surface structures is important for the amount of layering. For example, the flexible chain-like molecules tetradecane and hexadecane yield force curves oscillating more strongly with distance compared with the one for the flexible cyclic molecule cyclohexane. This can be seen in Fig. 3.1 where the separation between neighboring peaks in the force curves of the chain-like molecules is much narrower than for the cyclic one. This suggests preferential ordering of molecular symmetry axes with the surface. In

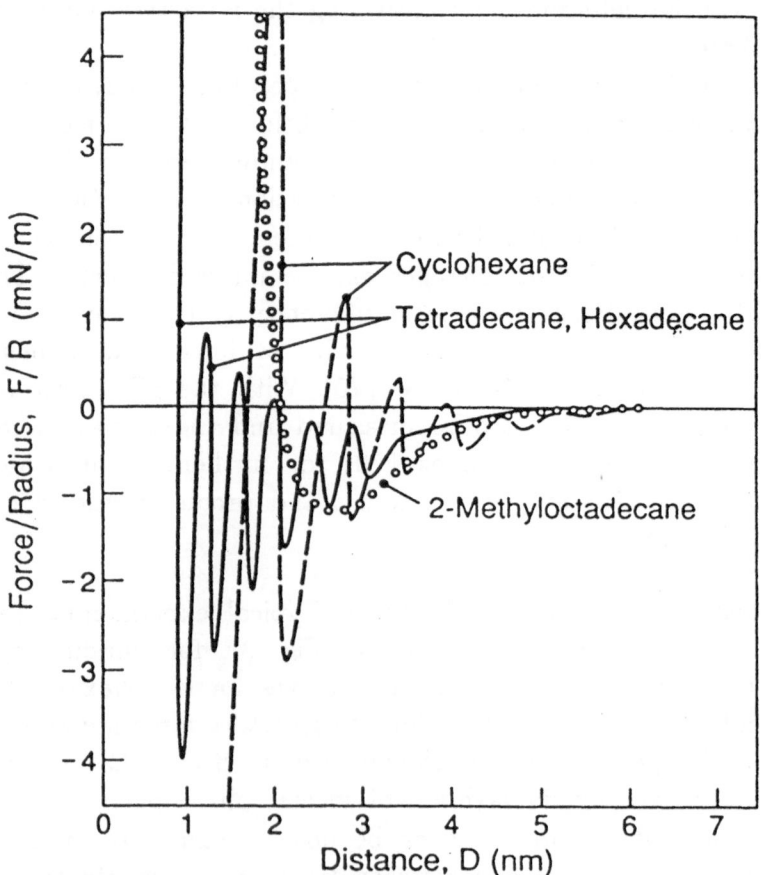

Fig. 3.1: Experimentally measured forces between mica surfaces for various organic vicinal phases (from [72]).

addition, the curves start at smaller surface separations in the case of chainlike molecules indicating more convenient packing between the surfaces. For the branched hydrocarbon 2–methyloctadecane there seems to be no significant layering because it lacks a molecular symmetry axis and can therefore not form discrete layers (see Chap. 13 in [70]).

A final comment applies to the abscissa scale. As can be seen in Fig. 3.1 the force curve damps out within $5\text{-}7\cdot10^{-9}$ m. This reflects the truly microscopic nature of structural effects one is dealing with in vicinal phases. The microscopic scales of observation involved experimentally render applications of computer simulations very appealing. The latter involve similar scales of observation but allow for a much more detailed picture because microscopic constituents forming vicinal phases are treated explicitly.

3.2.2 Total Internal Reflection Microscopy

TIRM is a technique by which forces between colloidal particles of radius $1\text{-}10\,\mu$m and a planar surface can be measured [68,77]. These forces are usually rather weak and of the order of O $(10^{-15}\,N)$. If a colloidal particle is placed in a container with some solvent, it will move downward on account of gravity. Sooner or later it will come to some sort of mechanical equilibrium at a distance $\langle h \rangle$ from the surface beneath as soon as repulsive forces between particle and surface balance gravitational forces.

The surface is usually made of glass [68,77] with good permeability for laser light. The laser beam is directed at the particle from below so that $\langle h \rangle$ can be deduced from the intensity of the reflected beam. However, colloidal particles are subject to rapidly changing forces exerted by solvent molecules between them and the glass surface. These fluctuating forces cause colloidal particles to exhibit Brownian motion. As time passes there will be a whole distribution of separations h around $\langle h \rangle$ and associated with it a change in intensity of the reflected light. From the change in intensity it is possible to deduce the force law for the colloidal particle's interaction with the surface.

The dynamical process behind TIRM bears some resemblance of the situation met in isostress-isostrain ensemble MC calculations to be discussed in detail in Sect. 4.3.2. As will be explained there, the thickness of films between plane parallel surfaces changes on account of density fluctuations in the film. The spatially fixed lower surface may be taken as being apropos of the TIRM glass surface while the upper moving surface behaves qualitatively like the colloidal particle hovering over the glass surface. Given the sizes of colloidal particles in the experiment their spherical shape may be well approximated locally by a planar surface over the length scale $(10\text{-}100\,\overset{\circ}{A})$ of a corresponding simulation system. In the isostress-isostrain ensemble MC calculations a similar distribution of h around $\langle h \rangle$ can be obtained. From its width the compressibility of the vicinal phase can be deduced which will be shown in Sect. 4.3.2 to be important in the context of phase transitions in vicinal phases.

3.2.3   The Atomic Force Microscope

The AFM is comparable to the SFA with the exceptions of greater sensitivity and one of the cylinders being replaced by a small tip [69,78]. Sensitivity of the force measurement is between $10^{-9}\text{-}10^{-10}\,\mathrm{N}$ and hence 1-2 orders of magnitude larger than for the SFA. Tip radii range from one atom to more than 1μm. Currently there is more emphasis on smaller tips which allow for direct measurements of forces between an individual atom and a surface or between two individual atoms.

To measure such small forces very sophisticated technology is required. This technology provides extremely small force sensing devices with spring stiffnesses as small as $0.5\,\mathrm{N\,m^{-1}}$ [70]. Using optical laser-based techniques, distances between tip and surface as small as 0.01 nm can be measured. This is one order of magnitude smaller than is reported for the SFA [70]. In [79] results of AFM experiments are reported in which long range electrostatic solvent forces are measured between a μm-quartz sphere and a flat surface in an aequous salt solution. The quartz sphere is attached to the tip of an AFM device and results are obtained out to a 60 nm separation between sphere and surface.

Possible sources of error in AFM experiments result from insufficient characterization and reproducibility of tip geometries. In addition the very fine tips are prone to elastic or plastic deformation during measurements.

This discussion of AFM concludes the brief survey of experimental techniques closest to the computer simulation studies reported here. It should be noted, however, that there is a whole wealth of other thermodynamic and spectroscopic experiments also revealing important information about vicinal phase properties. These techniques are comprehensively discussed in Israelachvili's book to which the interested reader is once again referred [70].

3.3 Models in Computer Simulation Studies

Although the previous discussion of key experiments demonstrates the remarkable precision with which forces in vicinal phases can be measured, vicinal phase structure cannot directly be studied. However, the interpretation of SFA data in Sect. 3.2.1 strongly suggests unique structural features of vicinal phases confined to spaces of molecular dimensions. For example, layering as suggested by SFA experiments is a phenomenon particularly distinguishing vicinal from bulk phases. However, layering as inferred from SFA must be regarded as a conjecture which remains to be verified in more rigorous terms.

Computer simulations can complement these experiments. They allow for a direct determination of vicinal phase structure due to the explicit treatment of individual particles.

However, before a computer simulation can be made a model needs to be specified. In the case of vicinal phases this amounts to specifying the geometry of the system and intermolecular potentials for interactions among vicinal phase particles and between vicinal phase particles and the confining walls.

There are basically two geometries one may consider. In Fig. 3.2 a schematic representation of a vicinal phase in slit-pore geometry is shown. Between two plane parallel square walls of side length $s_x^0 = s_y^0 = s$ vicinal phase particles are enclosed. The walls are a distance h apart. In the (x,y)-plane the system is assumed to be infinite. This is achieved in computer simulations with short range interaction potentials by applying periodic boundary conditions at the imaginary walls of the simulation cell in the (x,y)-plane [25].

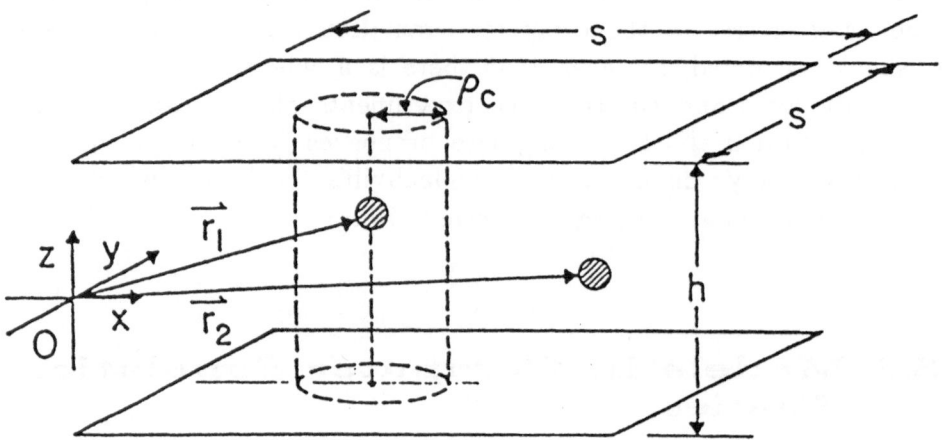

Fig. 3.2: Schematic of the model slit-pore, showing planes of solid layers, cutoff cylinder, and coordinate system employed; ρ_c is the cutoff radius.

A second representation of vicinal phases is provided by cylindrical pore geometry [80]. There vicinal particles are completely surrounded by confining walls in the (x,y)-plane while periodic boundary conditions are applied in z-direction to render the simulation system semiinfinite. However, there is little difference between the two geometries with regard to vicinal phase properties.

To proceed it is obvious from eqs. (2.1), (2.11), (2.14a)-(2.16) and (2.26)-(2.29) that one needs to specify the total potential energy $\Phi(\mathbf{r}^N)$. $\Phi(\mathbf{r}^N)$ can be written for vicinal phases in slit-pore geometry as a sum of three terms, namely

$$\Phi(\mathbf{r}^N) = \Phi_{pp}(\mathbf{r}^N) + \Phi_{pw}^{(1)}(\mathbf{r}^N;\mathbf{r}^{(1)}) + \Phi_{pw}^{(2)}(\mathbf{r}^N;\mathbf{r}^{(2)}) \qquad (3.1)$$

where the first term on the r.h.s. represents the contribution from interactions among vicinal phase particles. The two other terms represent contributions from the interaction between vicinal phase particles and lower, $^{(1)}$, and upper, $^{(2)}$, wall. Arguments $\mathbf{r}^{(1)}$ and $\mathbf{r}^{(2)}$ are used to indicate position and possible inherent structure of the walls. The precise form of $\Phi_{pw}^{(k)}(\mathbf{r}^N;\mathbf{r}^{(k)})$ depends on the particular wall model to be introduced subsequently in Sect. 3.3.1– 3.3.4.

Vicinal phase particles (i.e. atoms) are taken to be spherically symmetric without permanent multipoles. Their interactions are assumed to be pairwise additive so that

$$\Phi_{pp}(\mathbf{r}^N) = \sum_{i=1}^{N-1} \sum_{j=i+1}^{N} u(r_{ij}) \qquad (3.2)$$

where $r_{ij} = |\mathbf{r}_i - \mathbf{r}_j|$ and the intermolecular potential $u(r_{ij})$ is taken to be of the Lennard-Jones(12,6) form

$$u(r_{ij}) = 4\varepsilon \left[\left(\frac{\sigma}{r_{ij}} \right)^{12} - \left(\frac{\sigma}{r_{ij}} \right)^6 \right] \qquad (3.3)$$

Eq.(3.3) represents an effective pair potential because potential well depth ε and atomic "diameter" σ are usually obtained from experimental data like virial or transport coefficients and hence include implicitly contributions from many-body interactions [81]. True pair potentials as provided by ab-initio quantum mechanical calculations are only of limited use in computer simulations of dense fluids because they neglect the important many-body interactions. Eq. (3.3) also represents a typical short range potential because it decays faster than r_{ij}^{-3}. The rapid decay permits a potential cutoff ρ_c as indicated in Fig. 3.2: interactions between two vicinal atoms i and j are treated explicitly if particle j is located within a cylinder of radius ρ_c centered on reference particle i. If ρ_c is sufficiently large ($\gtrsim 3\sigma$), neglect of interactions beyond ρ_c causes a small error that may be corrected at the end of the simulation analytically [25,82].

Eqs. (3.2) and (3.3) describe interactions among vicinal particles regardless of wall models. These models need to be specified next.

3.3.1 Model I: Structured Soft Wall

Model I consists of a discrete distribution of $2N_s$ rigidly fixed interaction sites (i.e. surface atoms) across the plane of the confining walls (see Fig. 3.3). The inherent structure of the walls is taken to represent the (100)-plane of the face centered cubic (fcc) lattice in which rare gases solidify. Thus, $\{\mathbf{r}^{(1)}\}$ and $\{\mathbf{r}^{(2)}\}$ represent positions of surface atoms according to the fcc (100) structure. Coordinates of atoms in upper and lower walls are related by

$$x^{(2)} = x^{(1)} + \alpha l \tag{3.4a}$$

$$y^{(2)} = y^{(1)} + \alpha' l \tag{3.4b}$$

$$z^{(2)} = z^{(1)} + h \tag{3.4c}$$

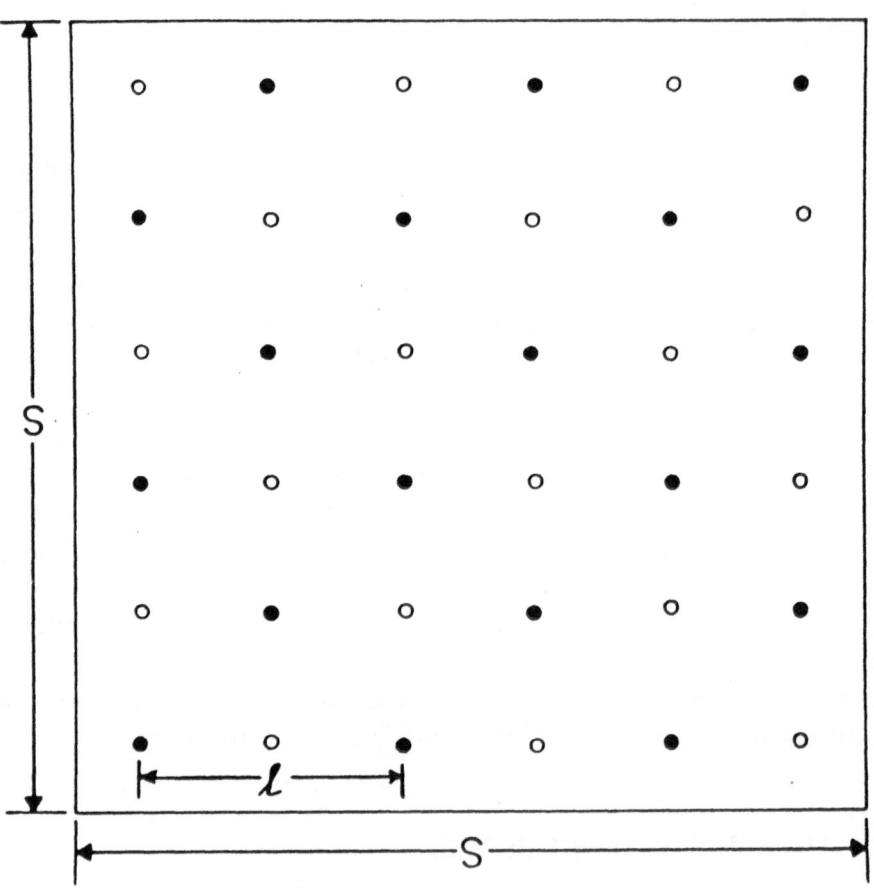

Fig. 3.3: Square unit cell of the (100) plane of the fcc lattice. Filled and open circles represent atoms in pore walls at $z = 0$ and $z = h$ (see eq. (3.4c)); l is the lattice constant.

where l is the lattice constant of the fcc lattice and $\{z^{(1)}\} = 0$. α and α' specify relative alignments of the two walls in the (x,y)-plane. They will be called "registry parameters" henceforth; α has been introduced already in eq. (2.22d) in conjunction with the displacement gradient tensor u It will be shown later in Sect. 3.4.2 that registry is important in the formation of solidlike vicinal structures; in Sec. 4.3 it will be demonstrated that these vicinal solids can be "melted" by applying shear strain to them. α is a measure of shear strain (see eq. (2.22d)). Note that the treatment in Sect. 2.3.2 does not include the application of shear strain along the y-axis in the spirit of eq. (3.4b). However, an extension of thermodynamical and statistical-physical considerations in Sect. 2.3.2 along this line would be straight-forward [64].

At α, $\alpha' = 0.0$ surface particles in both walls are exactly opposite of each other. This configuration will be called "in-registry". Fig. 3.3 shows an out-of-registry configuration for which $\alpha = 0.5$, $\alpha' = 0.0$.

Model I is called "soft" because surface and vicinal particles interact (predominantly) via a continuous intermolecular potential, namely the Lennard-Jones potential given in eq. (3.3). Thus,

$$\Phi_{pw}^{(k)}(\mathbf{r}^N; \mathbf{r}^{(k)}) = \sum_{i=1}^{N} u_{HS}(z_i) + \sum_{j=1}^{N_s} u(r_{ij}^{(k)})) \quad ; \quad k = 1, 2 \qquad (3.5a)$$

$$u_{HS}(z_i) = \begin{cases} \infty & , \ z_i \leq 0 \text{ or } z_i \geq h \\ 0 & , \ 0 < z_i < h \end{cases} \qquad (3.5b)$$

where $r_{ij} = |\mathbf{r}_i^{(k)} - \mathbf{r}_j^{(k)}|$ denotes the distance between vicinal atom i and surface atom j in wall k. $u_{HS}(z_i)$ represents an infinitely strong repulsive background potential. Due to its range compared with the Lennard-Jones potential this "hard-core" potential does not come into effect. It is introduced merely to define rigorously the vicinal phase volume V. The cutoff criterion introduced in Sect. 3.3 is applied here, too.

3.3.2 Model II: Structureless Soft Wall

To study the influence of discrete wall structure, model II is deduced from model I. Model II is obtained by "smearing" each of the N_s surface atoms over its respective plane. Mathematically this is achieved by integrating $u(r_{ij}^{(k)})$ over $x_j^{(k)}$ and $y_j^{(k)}$ while holding r_i fixed. The result is a mean field approximation to model I given by

$$u'(z_i^{(k)}) = 2\pi\varepsilon d_s \left[\frac{2}{5} \left(\frac{\sigma}{z_i^{(k)}} \right)^{10} - \left(\frac{\sigma}{z_i^{(k)}} \right)^4 \right] \tag{3.6}$$

where $d_s = N_s \sigma^2 / s^2$ is the surface density and $z_i^{(k)} = z_i - z^{(k)}$ represents the distance of vicinal atom i from wall k in z-direction.

By analogy with eq. (3.5a) one has

$$\Phi_{pw}^{(k)}(z^N; z^{(k)}) = \sum_{i=1}^{N} u_{HS}(z_i) + u'(z_i^{(k)}) \tag{3.7}$$

where the same comments regarding $u_{HS}(z_i)$ apply (see Sect. 3.3.1). Comparison of eqs. (3.5a) and (3.7) indicates that model II is computationally less demanding. While only single sums of N elements need to be computed here double sums of NN_s elements need to be evaluated for model I.

3.3.3 Model III: Structureless Hard Wall

In conjunction with EMD simulations to be reported in Sect. 3.4.2 and 5.2.3 two other wall models are considered. Model III consists of specularly reflecting hard walls so that (see eq.(3.5b))

$$\Phi_{pw}^{(k)}(z^N; z^{(k)}) = \sum_{i=1}^{N} u_{HS}(z_i) \tag{3.8}$$

The discontinuous forces resulting from eq. (3.8) can be handled in EMD as outlined in Sect. 2.2.2.

According to remarks made in that section the numerical procedure requires the velocity with which vicinal atoms leave the wall after having hit it. It can easily be found from the criterion of energy conservation. Since the wall is perfectly smooth x- and y-component velocities do not change upon atom-wall interaction so that

$$v'_{i,x} = v_{i,x} \tag{3.9a}$$

$$v'_{i,y} = v_{i,y} \tag{3.9b}$$

where again primed quantities refer to postcollisional values (see Sect. 2.2.2). Since energy must be conserved upon collision the only choice for $v'_{i,z}$ complying with eqs. (3.9a) and (3.9b) is

$$v'_{i,z} = -v_{i,z} \tag{3.9c}$$

3.3.4 Model IV: Infinitesimally Rough Hard Wall

Model IV is closely related to model III since it also employs eq. (3.8). However, here the wall is not perfectly smooth but possesses subatomar (i.e. infinitesimal) roughness. In other words, atoms colliding with the wall are scattered back stochastically into the vicinal phase. This renders trajectories of colliding atoms non-deterministic but again the scheme discussed in Sect. 2.2.2 can be employed to integrate the equation of motion. As before for model III energy must be conserved upon collision but in addition velocities of scattered particles must be distributed according to the probability density

$$f_A(\vartheta, \varphi)\, d\Omega' = \cos\vartheta\, d\vartheta\, d\varphi / \pi \tag{3.10}$$

In [51] it is shown that both conditions are fulfilled when

$$v'_{i,x} = v_i \sin\vartheta \cos\varphi = v_i\left[1 - \xi\right]^{1/2} \cos(2\pi\eta) \tag{3.11a}$$

$$v'_{i,y} = v_i \sin\vartheta \sin\varphi = v_i \left[1 - \xi\right]^{1/2} \sin(2\pi\eta) \qquad (3.11b)$$

$$v'_{i,z} = \pm v_i \cos\vartheta = \pm v_i \xi^{1/2} \qquad (3.11c)$$

where $v_i = |\mathbf{v}_i|$ and the sign in eq. (3.11c) is taken to be the opposite one of the precollisional z-component of $\mathbf{v}_i$. ξ and η are two pseudo-random numbers distributed uniformly on the interval [0,1]. As in Sect. 3.3.3 primed quantities refer to postcollisional values.

3.4 Computer Simulation Results

Computer simulations can now be performed to investigate vicinal phase structure. For the sake of later comparison with experimental results the simulation should be performed under conditions resembling experimental situations to a reasonable extent.

For instance, in Sect. 3.2.1 it is explained that in the SFA the crossed-cylinder configuration is immersed in a large reservoir of bulk fluid. After being brought into the reservoir, the vicinal phase forms as bulk molecules flow in-between the cylinder surfaces until bulk and vicinal phase have come to thermodynamic equilibrium. Clearly, this condition is fulfilled if $\mu'=\mu''$, ' and " referring to bulk respectively vicinal phase. Grand canonical MC is ideally suited to comply with this situation because μ,V,T are fixed thermodynamic state parameters in this ensemble. However, before vicinal phase structure is discussed in detail in Sect. 3.4.2 the claim to realism of simulation results demands some justification especially in view of the apparent simplicity of the models introduced in the previous sections.

3.4.1 Connecting with Experiments: The Solvation Force

If one wishes to demonstrate the relevance of models to experimental "reality" one should obviously compute properties accessible to experiments as well. According to the considerations in Sect. 3.2 the total force exerted on either wall would be an obvious choice. From eq. (3.1) this quantity is defined as

$$\langle F_z^{(k)} \rangle = \left\langle -\frac{\partial \dot\Phi_{pw}^{(k)}}{\partial z} \right\rangle \quad ; \quad k = 1,2 \tag{3.12}$$

where $z = z_{ij}^{(k)}$, $z_i^{(k)}$ depending on which model is employed. By symmetry due to slit-pore geometry one also has

$$\langle F_z^{(1)} \rangle = -\langle F_z^{(2)} \rangle \tag{3.13}$$

which provides a useful internal consistency check for the computer simulation. Associated with $\langle F_z^{(k)} \rangle$ is the so-called "solvation force" f_s defined as

$$f_s \equiv \langle F_z^{(1)} \rangle / s^2 = -\langle F_z^{(2)} \rangle / s^2 \tag{3.14}$$

which is directly determined in the SFA experiment (see Fig. 3.1). f_s is actually the negative of the normal pressure on the wall [70].

Fig. 3.4a displays results for f_s as a function of wall separation h obtained for model I in grand canonical ensemble MC calculations (* denotes dimensionless units in terms of Lennard-Jones potential parameters ε and σ; see Table 3.1 and Appendix B in [25]). Since μ, T and $V = s^2 h$ are fixed for each thermodynamic state, every point on the curves in Fig. 3.4a represents a single MC calculation. Fig. 3.4a shows that f_s oscillates as h changes and that these oscillations damp out the larger h gets. As h increases f_s gradually loses its oscillatory character and will eventually assume a value corresponding to the (isotropic) bulk pressure P_{bulk} in the limit $h \to \infty$. The functional dependence of f_s on h is very similar to the experimental results displayed in Fig. 3.1 although there much more complex systems are considered (see Sect. 3.2.1).

The reader should note a minor difference between the plots in Figs. 3.1 and 3.4a. While $f_s \to P_{bulk}$ in the limit of large h, the curve in Fig. 3.1 tends to zero. Since only force differences are measured by the SFA (see Sect. 3.2.1), the quantity plotted in Fig. 3.1 is related to f_s by

$$\Pi = f_s - P_{bulk} \tag{3.15}$$

Π is known as "swelling pressure" and is of experimental interest in soil physics and chemistry [83].

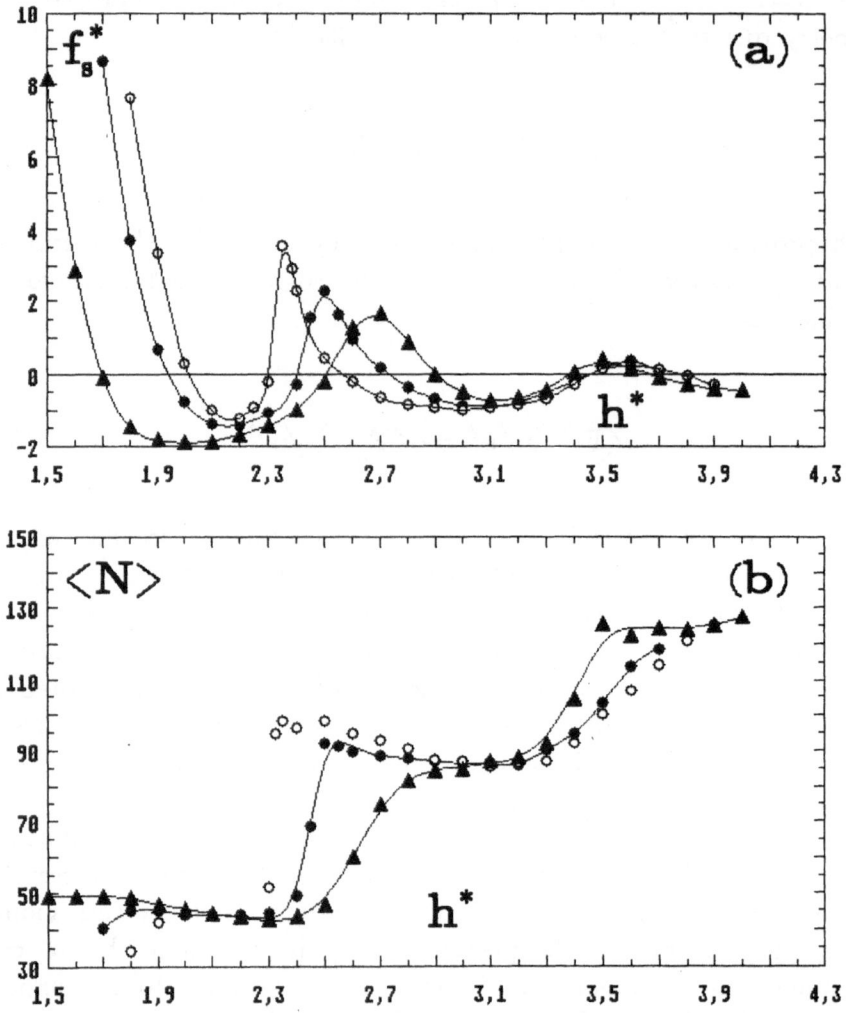

Fig. 3.4: **a:** The solvation force f_s as a function of interwall separation h for model I at $T^* = 1.0$, $-\mu^* = 12.0$ and $s^* = 7.9925$ from grand canonical ensemble MC calculations; (▲): $\alpha = 0.0$, (●): $\alpha = 0.3$, (○): $\alpha = 0.5$, (——): P_{bulk}. The full lines represent spline fits to the MC data to guide the eye. **b:** as **a** but for the average number of vicinal particles $\langle N \rangle$; due to the almost discontinuous change in the vicinity of $h^* \approx 2.3$ a similar spline fit is impossible to the curve at $\alpha = 0.5$.

Table 3.1: Reduced units of elementary physical quantities. $\varepsilon/k_B = 120\,K$, $\sigma = 3.4 \cdot 10^{-10}\,m$, $k_B = 1.3806 \cdot 10^{-23}\,J\,K^{-1}$, $m = 6.64 \cdot 10^{-26}\,kg$ (see eq.(3.3)). Other reduced units may be deduced from combinations of various entries listed in the table.

Quantity	In units of
Length	σ
Time	$(m\sigma^2/\varepsilon)^{1/2}$
Energy	ε
Temperature	ε/k_B

Magda et al. [84] also compute f_s as a function of h in EMD calculations based upon a model similar to model II. These authors find a qualitatively similar oscillatory dependence of f_s on h despite their use of structureless walls. However, since Magda et al. use EMD, N instead of μ is a fixed state parameter. They evade the use of an inappropriate ensemble by fixing N to values $\langle N \rangle$ previously obtained in grand canonical ensemble MC calculations by Snook and van Megen [85] who employ the same model. This procedure relies on the hypothesis of ensemble equivalence which has not been rigorously justified [86]. It can, however, often be justified heuristically and is employed as a working hypothesis here, too (see Chap. 5).

While the oscillatory character of f_s appears to be independent of the vicinal phase-wall potential and agrees qualitatively well with experimental results, details in the curves depend on the choice of Φ_{pw}. This is evident from Fig. 3.4a where results for f_s are presented for three different registries in x-direction. The curves are shifted with respect to one another along the abscissa; depths of minima and heights of maxima are affected by registry. At a particular h different numbers of vicinal atoms fit between the walls depending on α (see Fig. 3.4b).

This packing problem is somewhat similar to the experimental situation when the vicinal phase consists of branched as opposed to linear-chain hydrocarbons (see Sect. 3.2.1). Inappropriate molecular symmetry of the latter causes inhibition of layer formation. However, regardless of molecular details the degree of commensurability between surface and vicinal phase structures is responsible for the computationally as well as for the experimentally observable changes in the f_s curves. Due to their lack of molecular structure models II-IV do not allow for similar effects.

3.4.2 Vicinal Phase Structure

Layering in vicinal phases. To understand the origin of the previously discussed oscillatory dependence of f_s on h one needs to look at vicinal structure in detail. Computationally this can be done in principle in terms of local density $\rho^{(1)}(\mathbf{r})$. As pointed out in [82], in general this function depends on $\mathbf{r}$. However, to analyze $\rho^{(1)}(\mathbf{r})$ one would really need a four-dimensional representation which is hard to imagine or suitable cuts along certain axes. In any case, the amount of information would be overwhelming and hardly useful. In addition, from a computational point of view satisfactory statistical accuracy becomes a problem for a function of three independent variables. However, it seems reasonable to assume that structural effects will be most significant along the z-axis for slit-pore geometry. Thus, instead of $\rho^{(1)}(\mathbf{r})$

$$\rho^{(1)}(z) = \frac{\langle N(z) \rangle}{s^2 \, \Delta z} \tag{3.16}$$

will be analyzed here. In eq. (3.16) $\langle N(z) \rangle$ denotes the average number of particles in a small "slice" of volume $s^2 \Delta z$ centered on z. Fig. 3.5 displays results for $\rho^{(1)}(z)$ at various distances between the walls for model I. The curves are obtained in grand canonical ensemble MC calculations [87]. Panels a to g give clear evidence of layering in the vicinal phase.

Layering is, perhaps, first reported in [88]. Fig. 3.5 shows that the number of layers increases as the slit-pore widens due to the increase in space accessible to vicinal atoms.

Panel b in Fig. 3.5 demonstrates the effect of wall alignment. For the in-registry configuration only one layer of fluid fits between the walls while two layers can be accomodated at $\alpha = 0.5$. Panels a to e also show that an odd number of discrete layers forms with the walls in registry and an even number with the walls out of registry. The space between two discrete layers in panel b appears to be

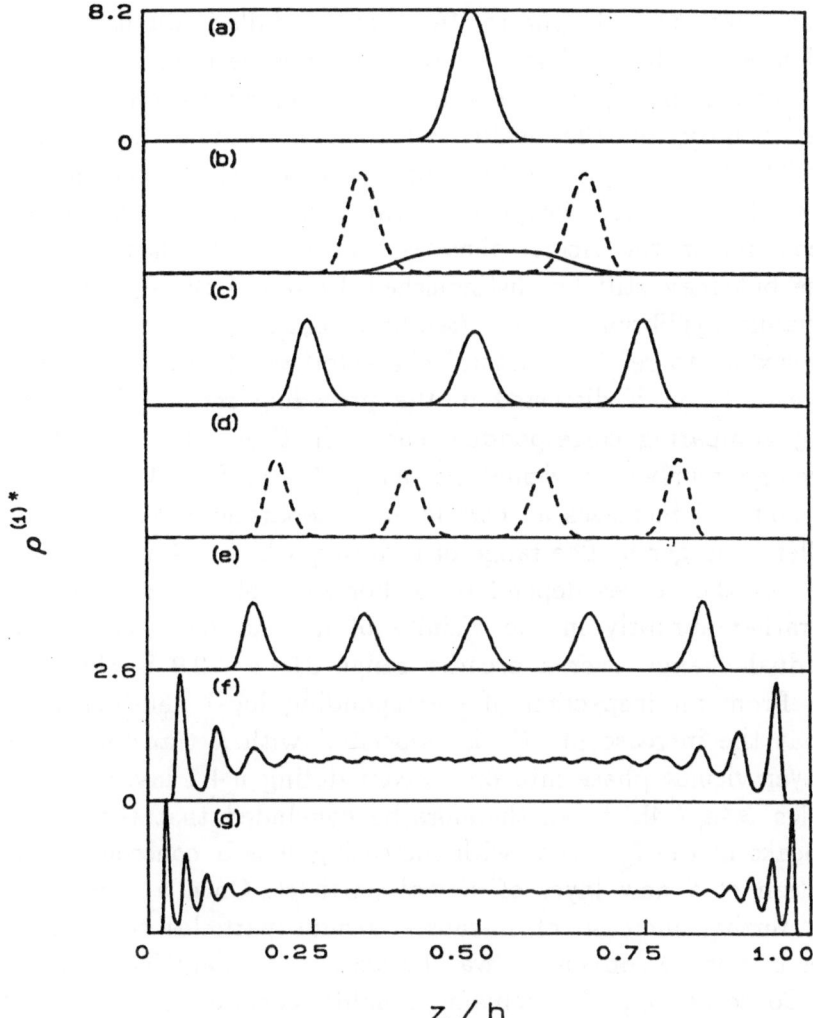

Fig. 3.5: Local density profiles from grand canonical ensemble MC calculations for model I ($T^* = 1.0$, $-\mu^* = 9.50$); **a**: $h^* = 1.37$, **b**: $h^* = 2.15$, **c**: $h^* = 3.05$, **d**: $h^* = 3.90$, **e**: $h^* = 4.90$, **f**: $h^* = 16.50$, **g**: $h^* = 30.00$. Full lines: $\alpha = 0.00$, broken lines: $\alpha = 0.50$.

inaccessible to vicinal atoms ($\rho^{(1)}(z) = 0$). From this rather strong spatial localization one may infer solidification of the vicinal phase. However, the information provided by $\rho^{(1)}(z)$ is not sufficient to establish this notion without doubt. For example, so far nothing can be said about the arrangement of particles within each discrete layer from which conclusions could be drawn concerning the structure of possibly solidlike vicinal phases. A more detailed discussion of solidification is therefore deferred to the following section.

Panels f and g in Fig. 3.5 show that layering occurs only in the immediate vicinity of the walls. Only five discrete layers are detectable. The remaining midsection appears to be rather homogeneous ($\rho^{(1)}(z) \approx$ const.) although wiggles are still visible at $h^* = 16.5$. In the homogeneous region the vicinal phase is isotropic and behaves like a bulk phase but may still be distinguished from a bulk reservoir in thermodynamic equilibrium with it (see Sect. 5.2.2).

It is rewarding to relate structural characteristics to the oscillatory dependence of f_s on h discussed in the preceding section. This can be done by comparing corresponding curves in Figs. 3.4a, 3.4b. Plots of the average number of vicinal particles $\langle N \rangle$ in Fig. 3.4b exhibit rather pronounced increases at certain wall separations as one goes from smaller to larger h. The range of h during which $\langle N(h) \rangle$ changes and details of the curves depend on α. For example, at $\alpha = 0.5$ $\langle N \rangle$ increases rather abruptly in the vicinity of $h^* \approx 2.3$ whereas only a rather gradual change over a broader range $2.5 \lesssim h^* \lesssim 2.9$ is observed at $\alpha = 0.0$. From an inspection of corresponding local densities it is evident that the increase of $\langle N \rangle$ is associated with a transformation of an n-layer vicinal phase into one accomodating $n + 1$ layers.

From Figs. 3.4a, 3.4b it can therefore be concluded that the occurrence of peaks in the f_s curves with increasing h is a consequence of the formation of a new layer of vicinal particles. Once a new layer is formed, further increase of h causes vicinal particles to arrange themselves in a more convenient way because more space is accessible to them. Consequently, $\langle N \rangle$ remains roughly constant or decreases slightly depending on the registry (see Fig. 3.4b) and the number of layers remains constant. The structural rearrangement correlates with a decrease of f_s (i.e. release of normal pressure). However, the increase of h is not yet sufficient to accomodate another layer of particles. Thus, f_s continues to drop until h is again large enough so that another layer can "pop" in.

If h is sufficiently large so that layering occurs only in the immediate vicinity of the walls, further increase of h leads to a rather gradual increase of the amount of matter in the vicinal phase. In this regime f_s loses its oscillatory structure and assumes a constant value corresponding to P_{bulk}. This can be understood from panels f and g in Fig. 3.5 because the bulklike homogeneous portion of the vicinal phase dominates the more the larger h gets. Since $\rho^{(1)}(z) \approx \rho_{bulk} = $ const.

one has to expect $f_s = P_{bulk}$ in the limit of large h as is observed in Figs. 3.1 and 3.4.

Layering similar to the oscillatory dependence of f_s on h is qualitatively independent of the nature of the walls. It is observed for soft structured [82] and unstructured walls [84] as well as for models III and IV [51]. Plots of $\rho^{(1)}(z)$ for the latter two in Fig. 3.6 show that the curves fall exactly on top of each other. Layering is qualita-

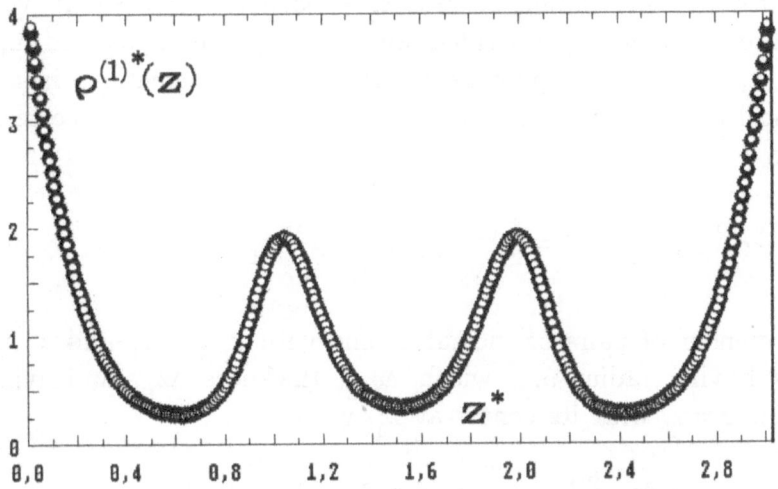

Fig. 3.6: Local density profile $\rho(z)$ for models III (●) and IV (○) from EMD calculations at T = 1.02, n = Nσ^3/V = 1.037.

tively similar to model I with the exception that the vicinal phases can wet the walls due to the vanishing range of Φ_{pw}.

From a more general perspective $\rho^{(1)}$ can be viewed as a cross pair correlation function $g_{12}^{(2)}$ for a binary mixture of gigantic spherical Brownian particles (i.e. the walls; component 1) and tiny solvent molecules (i.e. vicinal atoms; component 2). Peaks in $\rho^{(1)}(z)$ can then be interpreted as shells of neighbors around the Brownian particle so that from this perspective vicinal phase structure bears resemblance of bulk phase structure (see also Sect. 3.5.1).

Solidification of Vicinal Phases.

The In-Plane Pair Correlation Function. In the previous section the plots in panels a-e of Fig. 3.5 suggest that vicinal phases can solidify. To analyze this effect in more detail, structure has to be analyzed in somewhat more sophisticated terms.

This is achieved by investigating lateral in addition to normal structure. Quantitatively a discussion of lateral structure requires the next higher correlation function, namely $\rho^{(2)}(\mathbf{r}_1, \mathbf{r}_2)$ which represents the probability of finding a reference particle in a small volume element $d\mathbf{r}_1$ around $\mathbf{r}_1$ provided another particle is located in $d\mathbf{r}_2$ around $\mathbf{r}_2$ [8]. For slit-pore geometry it is convenient to introduce cylindrical coordinates and to define $\rho^{(2)}$ in terms of a pair correlation function $g^{(2)}$ as

$$\rho^{(2)}(z_1, \rho_{12}, z_{12}) = g^{(2)}(z_1, \rho_{12}, z_{12}) \rho^{(1)}(z_1) \rho^{(1)}(z_2) \qquad (3.17)$$

The number of pairs of vicinal atoms with $\mathbf{r}_1$ in $\Delta\mathbf{r}_1$ and $\mathbf{r}_2$ in an annulus having radius ρ_{12}, width $\Delta\rho_{12}$, thickness Δz_{12} and lying in the plane $z=z_2$ with its center at x_1, y_1 is

$$n = \rho^{(2)}(z_1, \rho_{12}, z_{12}) \Delta\mathbf{r}_1 \, 2\pi\rho_{12} \Delta\rho_{12} \Delta z_{12} \qquad (3.18)$$

Eq. (3.18) can be reexpressed as

$$n = \langle N(z_1, \rho_{12}, z_{12}) \rangle \rho^{(1)}(z_1) \Delta\mathbf{r}_1 \qquad (3.19)$$

where $\langle N(z_1, \rho_{12}, z_{12}) \rangle$ is the expected number of vicinal atoms in the above annulus with its center at (x_1, y_1, z_2) and reference atom 1 at (x_1, y_1, z_1). $\langle N(z_1, \rho_{12}, z_{12}) \rangle$ is computed by considering each vicinal atom in a layer of thickness Δz_1 about $z=z_1$ as a reference, counting the atoms in its annulus and averaging over all atoms in the layer about $z=z_1$ [82]. Restricting the analysis to in-plane distributions $(z=z_1)$ combination of eqs. (3.17)-(3.19) yields after simple algebraic manipulations

$$g^{(2)}(z_1, \rho_{12}) = \langle N(z_1, \rho_{12}) \rangle / (2\pi\rho_{12} \Delta\rho_{12} \Delta z_{12} \rho^{(1)}(z_1)) \qquad (3.20)$$

where $z_{12}=0$ has been dropped from the argument list to simplify notation.

The Role of Wall Structure. Eq. (3.20) is invoked to analyze lateral structure of vicinal phases. Their thermodynamic states pertain to data for $\rho^{(1)}(z)$ in Fig. 3.5. The corresponding plots in panels (a)-(e) in Fig. 3.7 give clear evidence of long range spatial order in vicinal phases for cases where $\rho^{(1)}(z)$ in Fig. 3.5 exhibits isolated discrete peaks.

Panel (b) in Fig. 3.7 demonstrates the effect of registry. While the out-of-registry configuration ($\alpha=0.5$) accomodates two discrete layers with long range lateral order, the in-registry configuration supports only one rather broad layer lacking significant long range order. This latter structure is typical of liquidlike Lennard-Jones fluids [52].

Long range order in general is indicative of highly ordered, solid-like structures. From peak positions in panels (a)-(e) of Fig. 3.7 it can be deduced that in the present case the vicinal phase forms the fcc (100) structure, i.e. it "freezes" epitaxially. Epitaxial formation of solidlike structures explains why only odd numbers of solid layers occur at $\alpha=0.0$ whereas only even numbers are observed at $\alpha=0.5$. The solidlike structure weakens as h increases which is particularly apparent from the characteristic double peak structure in $g^{(2)}$ around $\rho_{12}^* \approx 2.5$ in panel (a) of Fig. 3.7. The distinct double peak in panel (a) transforms gradually into a weak shoulder (panel (e)) indicating loss of long range order.

Plots of $g^{(2)}$ in Fig. 3.8 demonstrate that the liquid-solid transition occurs rather abruptly. While the vicinal phase at $h^*=2.05$ exhibits still liquidlike lateral structure, at $h^*=2.15$ long range solidlike order is already present. Further increase of h causes the solidlike vicinal phase to "melt" again for $h^*>2.8$ until solidification occurs again at $h^*=3.9$ (see Fig. 7 in [82]).

Abruptness of the phase transition is also reflected by the plots in Fig. 3.4b. For example, at $\alpha=0.5$ $\langle N \rangle$ increases in an almost discontinuous manner in the vicinity of $h^* \approx 2.3$ (Note the difference in μ between Figs. 3.4 and 3.8). From an analysis of structural changes in terms of $\rho^{(1)}$ and $g^{(2)}$ associated with the rapid change of $\langle N \rangle$, a liquid-solid phase transition is ascertained. If the registry changes toward $\alpha=0.0$ where the formation of a two-layer solidlike vicinal phase is prohibited, $\langle N \rangle$ changes more gradually upon formation of a second layer (see plots for $\alpha=0.3, 0.0$ in Fig. 3.4b).

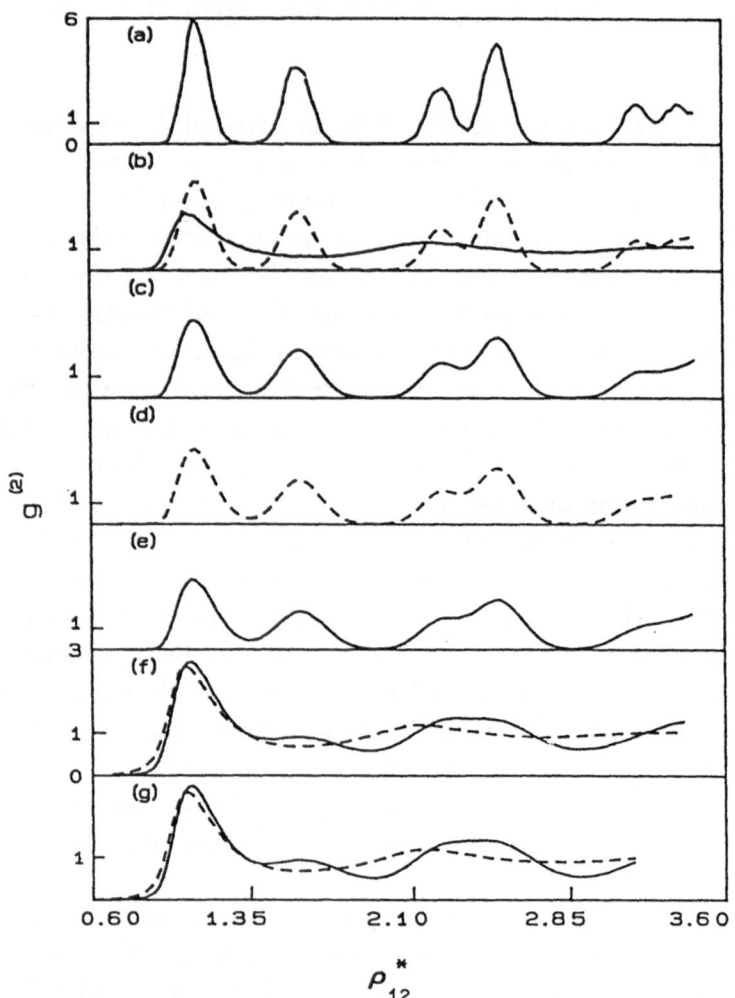

Fig. 3.7: In-plane pair correlation functions $g^{(2)}$ corresponding to thermodynamic states in Fig. 3.5. In panels (f) and (g) solid lines correspond to contact layer and dashed lines correspond to first inner layer.

From the plots in Fig. 3.4b it is also evident that "melting" upon increasing h is characterized by concomitant drainage. For instance, at $\alpha = 0.0$ where a one-layer solidlike vicinal phase forms epitaxially at $h^* = 1.5$, a small negative slope of $\langle N(h) \rangle$ is detected in the range $1.7 < h^* < 2.3$; at $h^* = 2.3$ the vicinal phase no longer exhibits solidlike

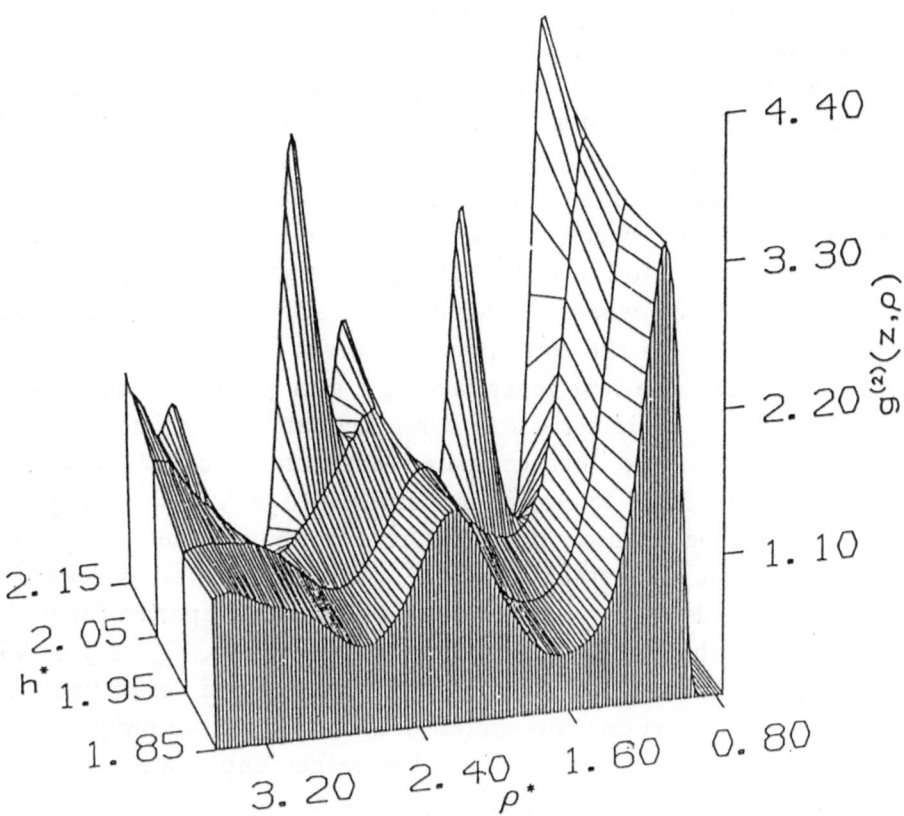

Fig. 3.8: Dependence of the in-plane pair correlation function $g^{(2)}$ on h^*; the argument z of $g^{(2)}$ denotes the vicinal layer bounded by planes $z^* = 0.6$ and $z^* = h^*/2$. Results are obtained in grand canonical ensemble MC calculations at $T^* = 1.0$, $-\mu^* = 9.10$ (see [82]).

structural characteristics. Similar observations are made for a two-layer vicinal phase ($\alpha = 0.5$) in the range $2.4 < h^* < 3.2$.

Formation of solidlike vicinal phases is not a triviality, even though the structures reported here are commensurate with the inherent structure of the walls. For instance, as is pointed out in [82] solidlike phases do not occur in the bulk at the same μ and T even if special care is taken to choose V so that a bulk solid would be geometrically possible. In addition, h does not bear an exact relation to l so that solidlike vicinal phases are distorted compared with bulk crystals. Thus, the walls act like templates in the formation of solidlike vicinal phases. This notion is corroborated by results for model II for which solidlike phases have never been observed for the same thermodynamic states (see Fig. 8 in [82]).

Formation of solidlike structures in vicinal phases was first observed in [82]. As can be seen there it is restricted to a certain range of h values. This range depends on a number of parameters such as μ, T and α. If h is not within an appropriate range only vestiges of solidlike structures can be detected in the contact layer (i.e. the layer closest to the walls). This is illustrated by plots in panels (f) and (g) of Fig. 3.7 where $g^{(2)}$ for contact layers exhibits weak peaks at positions at which the corresponding curves in panels (a)-(e) have distinct maxima. The first inner layer $g^{(2)}$'s in panels (f) and (g) of Fig. 3.7 are again characteristic of dense Lennard-Jones fluids.

Toxvaerd [89] who first compared structured fcc (100) and unstructured walls, failed to realize the importance of h in the formation of solidlike vicinal phases. His EMD results for a system of $h^* \approx 12$ led to the conclusion that only the contact layer is structurally affected by details of Φ_{pw}. Although this notion is supported for fairly wide pores by the results presented in panels (f) and (g) of Fig. 3.7, the previous discussion shows that Toxvaerd's findings cannot be extrapolated to very narrow pores. Results for self-diffusion to be presented in Sect. 5.2.3 also indicate the rather unphysical nature of structureless walls even if h gets very large so that effects on vicinal phase structure become marginal.

3.4.3   Results for Other Model Systems

It should be noted at this point that computer simulations of vicinal phases are not restricted to the models discussed so far.

For example, Heffelfinger et al. [90] investigate binary mixtures of argon and krypton in narrow cylindrical pores by means of EMD simulations. Argon and krypton are modelled as Lennard-Jones fluids for which the potentials are cut and shifted at some distance to avoid longer range interactions. The walls are chosen to represent carbon dioxide. The authors focus on capillary condensation and locate the transition point of two-phase coexistence.

Lennard-Jones mixtures are also investigated by Sokolowski and Fischer by means of constant-temperature MD simulations [91]. Potential parameters in their study are chosen to represent argon-krypton mixtures in a (structureless) carbon pore. The authors are mainly interested in vicinal phase structure and compare their simulation results with results of the Meister-Kroll-Groot version of density functional theory [92,93] (see Sect. 3.5.2). Sokolowski and Fischer find that density functional theory describes vicinal phase structure in accord with simulation data at three different temperatures in the range $1.0 \leq T^{*} \leq 2.0$.

An important extension of work based upon spherically symmetric potentials is concerned with long range electrostatic forces. van Megen and Snook [94] use grand canonical ensemble MC calculations to study the electric double layer. A version of grand canonical ensemble MC applicable to electrolyte solutions is developed in [95]. van Megen and Snook's model has slit-pore geometry with a discrete surface charge distribution. The electrolyte is represented by the restricted primitive model: ions are assumed to be singly charged hard spheres immersed in a continuum (the solvent) of dielectric constant ε. An important result of their study is the necessity of corrections due to the long range nature of the electrostatic potential. Especially in the vicinity of the walls such corrections are required. They can be incorporated by letting each particle interact with a continuous charge distribution outside the simulation cell. Results based upon this correction method indicate good agreement with the Gouy-Chapman theory up to charge densities of $0.1 Cm^{-2}$. At higher densities qualitative differences between this theory and numerical data are detected. These are ascribed to the neglect of finite particle extension in the Gouy-Chapman theory.

Torrie and Valleau [96] study the same model and obtain similar results. However, at high ion concentrations and surface charge densities counterions appear to be packed closely at the surface and begin to form layered structures. This results in a large electrostatic potential drop but only slight charge oscillations are observed in the solution.

Lee et al. [97,98] investigate vicinal phases in slit-pore geometry composed of Stockmayer molecules. They use EMD to study the effect of electric fields on vicinal phase structure over a wide range of field strengths. Long range dipolar forces are incorporated via Ewald sums. A special formulation for Ewald sums in confined systems with cylindrical symmetry can be found in [99]. A point of criticism that might apply to Lee et al.'s work concerns the nature of the thermodynamic states considered. Since the authors have no information about chemical potentials it cannot be decided with certainty whether the bulk phase in equilibrium with the vicinal phase is in a thermodynamically stable state. Various methods to compute μ in Stockmayer bulk phases are compared by Han et al. [100].

A large number of simulations are concerned with structure and to some extent with dynamical properties of vicinal water [101-113]. Perhaps the most sophisticated simulation of vicinal water at present was conducted by Delville [114] who employed grand canonical ensemble MC to investigate a clay-water interface. His model consists of a structured wall resembling montmorillonite and the MCY model for water-water interactions [115]. The clay-water interaction is represented as a sum of pairwise additive atomic interactions whose parameters are determined by fitting to the potential energy hypersurface computed quantum-mechanically by the self-consistent field procedure in the MNDO approximation [116]. "10 days on an Alliant superminicomputer" [114] are needed to carry out the computation at one state point. The results indicate that around the middle of the pore of width $h = 20 \overset{\circ}{A}$ vicinal water density corresponds to the density of bulk water. This is rather surprising in the light of very extensive experimental work by Drost-Hansen who found that surfaces perturb water over distances of 100-300 $\overset{\circ}{A}$ [75,76]. Clearly, properties of vicinal water are an important subject especially with regard to their influence on biological processes in living cells [78].

Vicinal phases composed of larger molecules than water are also reported. Honeycutt and Thirumalai conduct a study of static properties of polymers in porous media [117]. The porous medium is simulated by a site percolation model in which various sites are randomly occupied. A freely jointed chain is allowed to move in continuous space between the obstacles, i.e. the sites of the percolation model. Correlation functions such as the distribution of end-to-end distances and density profiles of monomers with respect to the center of mass of the chain are computed in MC calculations. It is found that the polymer molecule undergoes shape fluctuations in the random environment and arguments are presented which suggest that these fluctuations are important for transport mechanisms for polymers in random media.

3.5 Numerical Approaches Besides Computer Simulations

3.5.1 The Born-Green-Yvon Equation

The discussion in Sect. 3.4.2 demonstrates that local density $\rho^{(1)}$ is the simplest quantity revealing a lot of useful information about vicinal phase structure. Besides computer simulations discussed so far, numerical solutions of integral equations provide ways to determine $\rho^{(1)}$.

Such an equation may be derived from Liouville's equation which describes the time evolution of a classical N-particle-system in phase space Ω [8] via a phase space distribution function $f^{(N)}(\mathbf{p}^N, \mathbf{q}^N; t)$. $f^{(N)}(\mathbf{p}^N, \mathbf{q}^N; t)$ is the time-dependent analogue of $f_0^{(N)}$ introduced in Sect. 2.3. Unfortunately, Liouville's equation cannot be solved analytically in most cases due to the dimensionality of Ω. However, the equation is still useful in certain cases where the full information about the entire set of N particles is not really needed. Time evolution of the N-particle-system may then be approximated by a representative subsystem of n particles $(n < N)$. In those cases $f^{(N)}$ can be replaced by a reduced distribution function $f^{(n)}$ which can be obtained from $f^{(N)}$ by an integration over all coordinates and momenta of the remaining $(N - n)$ particles.

Integrating Liouville's equation over all those coordinates and momenta leads to the so-called *Born-Bogolyubov-Green-Kirkwood-Yvon* (BBGKY)-hierarchy of integro-differential equations in which a reduced distribution function of order n is expressed in terms of the (equally unknown) distribution function of order $(n+1)$. Therefore, solutions of the BBGKY-hierarchy at any level depend critically on the choice of closure.

If one is solely interested in structural properties one may assume a Maxwell-Boltzmann distribution of momenta and integrate the BBGKY-hierarchy over momentum subspace analytically to obtain a similar hierarchy of integro-differential equations for correlation functions $g^{(n)}(\mathbf{r}_1, \mathbf{r}_2, \ldots, \mathbf{r}_n)$ [8]. For $n=2$ a sensible closure of this hierarchy is provided by Kirkwood's superposition approximation [19] (see also Chap. 1) so that the *Born-Green-Yvon* (BGY) equation for the pair correlation function $g^{(2)}$ may be derived (for details, see Sect. 13-3, 13-4 in [8]).

Based upon the BGY equation, Fischer and Methfessel derive an integral equation for $\rho^{(1)}$[118]. Implicitly this approach interprets $\rho^{(1)}$ as a cross pair correlation function in a binary mixture composed of vicinal atoms and walls according to the ideas introduced earlier in Sect. 3.4.2 (see also [70]). However, additional assumptions are required to arrive at an equation for $\rho^{(1)}$.

First, a Weeks-Chandler-Andersen (WCA)-type [119] of potential split is necessary [120]. This leads to an additive separation of repulsive and attractive terms in the integral term of the BGY equation. For the longer range attractive part structure is neglected, i.e. $g^{(2)}(r) = 1$. The repulsive part is approximated by a hard sphere fluid with an effective hard sphere diameter [120]. The resulting expression involves the hard sphere pair correlation function at contact for the inhomogeneous vicinal phase. It is approximated by a pair correlation function at contact in a corresponding homogeneous bulk phase at some effective density. This bulk value can easily be obtained from the Carnahan-Starling hard sphere equation of state [121].

Based upon these approximations Fischer and Methfessel [118] arrive at

$$\nabla_1 \ln \rho^{(1)}(\mathbf{r}_1) = -\beta \nabla_1 u_{pw}(\mathbf{r}_1) + \int \delta(r_{12} - d) \frac{dr_{12}}{d\mathbf{r}_1} g_{bulk}^{HS}(d,\bar{n}) \rho^{(1)}(\mathbf{r}_2) \, d\mathbf{r}_2$$

$$-\nabla_1 \int \beta u^{(1)}(r_{12}) \rho^{(1)}(\mathbf{r}_2) \, d\mathbf{r}_2 \tag{3.21}$$

In eq. (3.21) u_{pw} is the particle-wall potential, d is the effective hard sphere diameter, $\bar{n}$ is the effective bulk density, g_{bulk}^{HS} is the bulk phase hard sphere pair correlation function and $u^{(1)}$ is the attractive potential in the WCA split. Eq. (3.21) can be solved iteratively in cylindrical coordinates for the unknown function $\rho^{(1)}$ provided some sensible starting value is known [118].

In a series of papers [122-126] eq. (3.21) is used to study vicinal phase structure and associated thermodynamic properties. Heinbuch and Fischer present $\rho^{(1)}$ from eq. (3.21) in comparison with EMD data. The agreement is qualitatively good but eq. (3.21) overestimates layering as one moves away from the walls; normal structure is underestimated in the contact layer [125], i. e. $\rho^{(1)}$ from eq. (3.21) is significantly lower compared with EMD data. Sokolowski and Fischer

[126] discuss the relation between the solution of eq. (3.21) and chemical potential. Their approach is important with respect to adsorption of fluids in pores. Sokolowski and Fischer also extended eq. (3.21) to treat multicomponent mixtures. Therefore, coarse grained densities are introduced to approximate contact values of hard sphere pair correlation functions. Results are in good agreement with MC data.

Based upon the BBGKY hierarchy Zhou and Stell derive an equation for $\rho^{(1)}$ which involves a superposition approximation of the triplet correlation function in terms of three pair correlation functions [127]. Their method predicts crystallization in contradiction to computer simulation results of [82]. Zhou and Stell also use a different approach based upon solutions of the Ornstein-Zernike equation. Invoking both hypernetted chain and Percus-Yevick closures, they present results in very good agreement with [82]. All three methods are applied to cylindrical pores as well [128] and are used to investigate microemulsions of oil in water [129].

Reference hypernetted chain theory is also used by Torrie et al. [130] to study the structure of hard sphere fluids next to curved surfaces of isolated macroions of varying diameters and surface charges.

3.5.2   Density Functional Theory

A second non-simulation numerical approach to $\rho^{(1)}$ is based upon density functional theory [131]. In this method one writes the grand potential ψ for a vicinal phase as

$$\psi[\rho^{(1)}(\mathbf{r})] = F[\rho^{(1)}(\mathbf{r})] + \int \left[v(\mathbf{r}) - \mu \right] \rho^{(1)}(\mathbf{r}) \, d\mathbf{r} \qquad (3.22)$$

in terms of intrinsic free energy F and chemical potential μ. $v(\mathbf{r})$ is an external attractive potential representing the walls. Since $\rho^{(1)}(\mathbf{r})$ is a spatially varying function both F and ψ become functionals of $\rho^{(1)}(\mathbf{r})$. Its equilibrium form can be obtained by making ψ stationary with respect to the density profile, i.e.

$$\frac{\delta \psi[\rho^{(1)}(\mathbf{r})]}{\delta \rho^{(1)}(\mathbf{r})} = 0 \qquad (3.23)$$

To solve eq. (3.23) several approximations are required. First, F is split in a WCA fashion into repulsive (index "r") and attractive (index "a") parts and is written as

$$F[\rho^{(1)}(\mathbf{r})] = F_r[\rho^{(1)}(\mathbf{r})] + F_a[\rho^{(1)}(\mathbf{r})] \qquad (3.24)$$

The attractive forces are usually treated in a mean field type of approach which ignores completely any correlations between particles. If u_a represents the attractive part of the intermolecular potential one may write F_a in the mean field form as

$$F_a = \frac{1}{2} \int\int d\mathbf{r}\, d\mathbf{r}'\, \rho^{(1)}(\mathbf{r})\, \rho^{(1)}(\mathbf{r}')\, u_a(|\mathbf{r} - \mathbf{r}'|) \qquad (3.25)$$

The repulsive part which determines vicinal phase structure is subject to various assumptions by which specific forms of density functional theory differ. In its simplest form based upon the local density approximation F_r is written as [132]

$$F_r = \int d\mathbf{r}\, f_{HS}[\rho^{(1)}(\mathbf{r})] \qquad (3.26)$$

where f_{HS} is the Helmholtz free energy density of a uniform hard sphere reference fluid of density ρ. This causes short range correlations to be excluded. Unfortunately, these short range correlations are responsible for the oscillatory form of $\rho^{(1)}$ in the vicinity of the walls. Consequently, $\rho^{(1)}$ obtained from eq. (3.23) does not exhibit oscillations if the local density approximation is invoked.

Several authors have, therefore, tried to improve the treatment of repulsive forces by introducing more sophisticated forms for F_r. In these approaches repulsive forces are treated in a non-local fashion [133-135]. These non-local forms of density functional theory are capable of describing the oscillatory behavior of $\rho^{(1)}$ in the vicinity of the walls in agreement with computer simulation results. However, it would be beyond the scope of this article to critically discuss merits and shortcomings of each individual theory. However, a numerically particularly simple approach [136] is the one by Kierlik and Rosinberg [137] who begin by splitting F_r into an ideal gas part

$$F_r^{id}[\rho^{(1)}(\mathbf{r})] = \beta^{-1}\int d\mathbf{r}\,\rho^{(1)}(\mathbf{r})\big[\ln\big(\Lambda^3\rho^{(1)}(\mathbf{r}) - 1\big)\big] \qquad (3.27)$$

and into an excess part for which they write

$$F_r^{ex}[\rho^{(1)}(\mathbf{r})] = \beta^{-1}\int d\mathbf{r}\,\Phi^{PY}\big(\rho_0(\mathbf{r}),\rho_1(\mathbf{r}),\rho_2(\mathbf{r}),\rho_3(\mathbf{r})\big) \qquad (3.28)$$

where Φ^{PY} results from the Percus-Yevick compressibility equation of state

$$\Phi^{PY} = -\rho_0\ln(1-\rho_1) + \rho_1\rho_2/(1-\rho_3) + \frac{1}{24\pi}\rho_2^3/(1-\rho_3)^2 \qquad (3.29)$$

The functions ρ_i, $i=0,.....,3$ are four weighted densities defined as

$$\rho_i(\mathbf{r}) = \int \rho_i(\mathbf{r}')\,w_i(|\mathbf{r}-\mathbf{r}'|)\,d\mathbf{r}' \qquad (3.30)$$

where the weighting functions w_i can be expressed in terms of the Dirac δ-function, its first and second derivative and the Heaviside step-function [136]. The coarse grained, weighted density approach by Kierlik and Rosinberg [137] is similar in spirit to other forms of density functional theory [138-141].

Density functional theory has also been extended to study binary mixtures of spherically symmetric molecules [90,142,143]. Tang et al. [144] have recently applied density functional theory to a dipolar hard sphere fluid in contact with neutral hard walls. They investigate position-orientation-dependent distribution functions. The results agree with computer simulation results by Levesque and Weis [145] for the same model at high densities. At low densities and high dipole moment density functional theory predicts a qualitatively different interfacial structure from that at high densities. A related model in which a dipolar vicinal phase is confined by charged walls has been studied by Rickayzen and Grimson [146].

It should be noted that density functional theory is potentially useful with regard to phase transitions in vicinal phases like capillary condensation or wetting transitions because it deals explicitly with the grand potential. A discussion of density functional theory in the context of phase transitions in vicinal phases is deferred to Chap. 4.

4 Phase Transitions in Vicinal Phases

4.1 Preliminary Remarks

The previous discussion highlights the unique structural characteristics of vicinal phases. For example, in Sect. 3.4.2 it is shown that vicinal phases can solidify provided certain prerequisites are met. Besides inherent wall structure interwall separation and registry are important to initiate solidification. In addition to these geometric factors thermodynamic conditions determine the nature of vicinal phases. For example, if at a given temperature the chemical potential is too low, the vicinal phase may not be able to accomodate enough atoms to form an ordered solidlike structure: only a rather disordered but dense and significantly layered liquidlike vicinal phase exists. At even lower chemical potentials only a thin layer of gaseous material may be adsorbed on the walls leaving most of the space in-between the walls empty.

Besides any two of these three states of matter phase transitions will occur depending on changes of thermodynamic or geometric conditions. However, due to the inhomogeneous and anisotropic nature of vicinal phases these phase transitions are unique in many aspects. For example, shear melting which is discussed in Sect. 4.3 turns out to be a second order liquid-solid phase transition driven by geometric changes. In the bulk, on the other hand, liquid-solid phase transitions cannot occur as second order but must occur as first order phase transitions [38]. A typical first order phase transition involving vicinal phases is capillary condensation. Capillary condensation bears close resemblance to bulk phase liquid-gas transitions. However, experimentally capillary condensation is almost inevitably accompanied by hysteresis, that is liquid-to-gas transitions occur at a different chemical potential (i.e. pressure) compared with gas-to-liquid transitions. Hysteresis of this sort is not observed in bulk phases.

4.2 Capillary Condensation

4.2.1 Sorption Experiments

In homogeneous bulk phases condensation is a well known phenomenon. It occurs on isotherms below the critical temperature if the pressure is raised above the saturated pressure of bulk gas. Condensation is a spontaneous, reversible process according to characteristics of first order phase transitions. A similar process is observed in porous media. If the chemical potential in a bulk reservoir connected to the porous medium is raised beyond a certain value the porous medium fills with fluid and an initially gaslike vicinal phase undergoes a transition to a liquidlike one [147-149].

Quantitatively, liquid-gas transitions in vicinal phases can be discussed in terms of the adsorbed (desorbed) amount of matter. The resulting adsorption (desorption) (i.e. "sorption") isotherms are characterized by a number of interesting features. For example, analogous to a bulk system a capillary critical temperature T_c^{cap} can be defined. In general, due to confinement (see Sect. 4.2.4) $T_c^{cap} < T_c^{bulk}$ (bulk critical temperature). Sorption isotherms for $T > T_c^{cap}$ are continuous and reversible: the pore fills up gradually with material from the bulk reservoir.

Below T_c^{cap} liquid-gas transitions are accompanied by discontinuous changes in the sorption isotherms in the vicinity of some critical chemical potential. However, subcritical (with respect to T_c^{cap}) experimental sorption isotherms are almost inevitably characterized by hysteresis, i.e. the location of the transition point depends on whether the transition is approached along the gas- or along the liquidlike branch of the sorption isotherm.

The precise form of the hysteresis loops depends on a number of features. For instance, the slope of the sorption isotherm in the loop is the larger the narrower the size distribution in the porous medium. For a perfectly sharp distribution (i.e. a uniform pore size) the hysteresis loop should be vertical at the transition points.

The extent of hysteresis depends also in many ways on thermodynamic conditions and varies from system to system. Sometimes hysteresis loops shrink in size and may even vanish as temperature increases [150]. In other cases loops may decrease if the temperature is lowered toward the bulk triple point [151]. However, shrinkage with increasing temperature seems to be the most commonly observed effect. Usually hysteresis vanishes for temperatures below T_c^{cap}. Therefore, Ball and Evans suggest to associate hysteresis with the onset of critical capillarity [132] which is predicted to occur in single pores.

An experiment furnishing this interpretation has been performed by Awschalom et al. [152]. Advances in glass-making technology permit these authors to manufacture cylindrical micropores having narrow and controllable size distributions. Mercury porosimetry measurements reveal size distributions for which about 96% of the pores have radii deviating as little as 5% from the average radius of the entire sample [152]. Awschalom et al. find clear evidence of hysteresis in sorption isotherms of oxygen and xenon [152]. Since the pores are virtually identical, sorption in each one of them must give rise to hysteresis independently and collective effects due to connectivity among pores play only a marginal role.

Unfortunately, most substrates employed in sorption experiments are far from having such narrow pore size distributions. Nevertheless their sorption isotherms exhibit hysteresis. In these cases hysteresis must be a result of a highly cooperative mechanism in which connectivity among pores and thermodynamic states in neighboring pores play vital roles [153]. This physical picture is closely related to effects like supercooling or -heating in bulk fluids.

4.2.2 Application of Density Functional Theory to Capillary Condensation

Density functional theory is ideally suited to investigate the origin of hysteresis theoretically. In essence density functional theory minimizes the grand potential ψ with respect to the density profile $\rho^{(1)}(\mathbf{r})$ (see Sect. 3.5.2). While this procedure yields $\rho^{(1)}(\mathbf{r})$ as a solution, it simultaneously allows one to determine directly the equilibrium value(s) of ψ. Thermodynamically stable states are characterized by unique solutions of the minimization procedure, whereas multiple solutions exist for metastable states.

Density functional theory is applied to capillary condensation by a number of groups [132,154-160]. According to the previously described experimental situation Ball and Evans investigate capillary condensation for essentially two different models of porous media. Their first model consists of a single pore [159] reminiscent of the experiment performed by Awschalom et al. [152]. The second model employs an interconnected network of pores differing in size [132]. This model was originally proposed by Mason [153]. The network model is more apropos of the commonly encountered experimental situation described at the end of Sect. 4.2.1.

Despite its shortcomings (see Sect. 3.5.2), in both cases the local density version of density functional theory is employed because of its simplicity. As pointed out by Ball and Evans [132] this approach can be justified on grounds of the apparent insensitivity of (integral) thermodynamic properties like ψ with respect to (local) structure.

In the single pore model minimizing ψ yields unique solutions for gas- and liquidlike branches of sorption isotherms outside the hysteresis loop. Inside this loop density functional theory gives two solutions, one belonging to the gas- the other one belonging to the liquidlike branch. One of these is the global minimum of ψ while the other one represents only a local minimum. Thus, it can be decided at any point which branch of the isotherm is thermodynamically stable and which one is only metastable. As the transition point is approached the two solutions for ψ become increasingly similar and identical at the phase transition (i.e. capillary coexistence). However, the jump from one branch of the isotherm to the other does not occur precisely at capillary coexistence; the system remains on either branch of the isotherm as long as metastable states are available. On the contrary, the phase transition occurs at points (i.e. chemical potentials) demarcating the hysteresis loop where only unique minima of ψ exist. So in the single pore model hysteresis arises from the existence of metastable thermodynamic states.

A somewhat more sophisticated model is also used by Ball and Evans [132]. Instead of a single pore they use a gaussian pore size distribution with parameters suggested by Saam and Cole for adsorption of helium on vycor [161]. Individual pores are assumed to be independent and behave as single pores of different radii, i.e. each pore has free access to the connected bulk reservoir. Again hysteresis arises on account of metastability. The size distribution mainly affects the steepness of the continuous sorption isotherms for the entire porous medium in the vicinity of the hysteresis loop.

However, some caution is advised in the interpretation of the above results. As Hill points out hysteresis in equilibrium statistical-physical *theories* is a result of various approximations to the true partition function and not an exact physical law [162]. This could easily apply to density functional theory because of the approximations involved. According to Hill hysteresis in experiments (as the one by Awschalom et al. [152]) should be a non-equilibrium phenomenon that cannot be properly understood within the restrictive framework of equilibrium statistical physics. Interestingly, Awschalom et al. interpret their results in terms of a hydrodynamic model proposed by Cole and Saam [161] which relates the thermodynamic state of the pore fluid to the behavior of long-wavelength modes as a function of thickness of the adsorbed layer.

In the network model an important new feature is introduced by connecting each pore to a certain number of neighboring pores. The "windows" through which pores are connected differ in size according to some distribution. As in the independent pore model pore sizes differ according to another distribution. Both distributions are related by the assumption that one of the windows has a radius equal to the pore radius.

From Ball and Evans' calculations it turns out that adsorption and desorption are quite different processes in the network [132]. On adsorption each pore behaves like an isolated pore in the independent model: the amount adsorbed depends solely on pore radius and chemical potential and can be determined via density functional theory. The jump in the adsorption isotherm for each individual pore is assumed to occur precisely at its capillary coexistence point. There are no metastable states involved [132].

On desorption, however, pore blocking arises because windows may be too small for material to "flow" through and smaller neighboring pores may still be filled themselves. So for chemical potentials just below capillary condensation there is virtually no desorption on account of the blocking effect. If the chemical potential is sufficiently lowered some critical value will eventually be reached at which material is desorbed from a significant fraction of pores. At this point the network empties abruptly in some sort of chain reaction so that desorption may be viewed as a percolation process [132]. It is important to note, that in this model hysteresis occurs on account of the blocking effect and is not caused by metastable thermodynamic states as in the single or in the independent pore model [132].

4.2.3 Hysteresis in Computer Simulations: A Critique

Ergodicity and Metastability. In principle computer simulations provide an alternate route to phase transitions. For example, employing versions of the Metropolis algorithm (see Sect. 2.3) one generates realizations of Markov chains so that the characteristic potential pertaining to the ensemble used is minimal. In other words, in MC (as well as in EMD) one simulates in principle thermo-dynamic equilibrium states provided the Markov chain is ergodic.

However, ergodicity is not automatically guaranteed. A useful and lucid discussion of ergodic problems associated with MC is presented by Wood (p. 115 in [58]). Markov chains are ergodic if any microstate having non-vanishing probability of occurrence can be reached in a finite number of steps from any other state having a non-vanishing probability of occurrence. In practice, however, the system may get trapped in a localized region of configuration space. For a Markov chain of finite length it may not be possible to reach any other region that has a significant probability of occurrence because the "bottlenecks" connecting the two are narrow and difficult to find (see also Fig. 2.1 in [25]). Certain computed properties may be non-negligibly influenced by these inadvertently neglected regions. Wood states that the ergodic problem "is not infrequently closely associated with size and perhaps the shape of the system and with the boundary conditions.". This comment should be borne in mind for the following discussion of computer simulation results showing hysteresis for capillary condensation.

Capillary condensation (and phase equilibria in general) can be studied by grand canonical ensemble MC because ψ is directly accessible in terms of simple "mechanical" properties like pressure (see, for instance, eq. (2.4.13) in [52]). For inhomogeneous systems in slit-pore geometry Henderson and van Swol derive a corresponding expression by considering the work done on a subsystem during the creation or destruction of part of it [163,164]. This leads to

$$\psi = -A \int_0^h P_T(z)\, dz = -\langle N \rangle \beta^{-1} + \frac{1}{2} \left\langle \sum_{i>j} \sum r_{ij}^{-1}(r_{ij}^2 - z_{ij}^2)\frac{d\Phi(r_{ij})}{dr_{ij}} \right\rangle \qquad (4.1)$$

where $P_T(z)$ denotes the transverse part of the pressure tensor and A is the area of the walls. This expression cannot be applied to cylindrical geometries [165].

Capillary condensation is investigated in a variety of systems by means of grand canonical ensemble MC [166-171] or other MC methods [172]. Snook and van Megen calculate sorption isotherms for single vicinal phases in slit-pore geometry composed of atoms and confined between unstructured walls [167]. Their isotherms (see Fig. 2 in [167]) are presented as smooth reversible curves even for temperatures below T_c^{cap}. For the single pore model studied by Snook and van Megen [167] density functional theory predicts hysteresis as a result of metastable thermodynamic states [159]. Snook and van Megen's sorption isotherms are apparently not affected by metastability.

Metastability is, however, observed in a number of other simulations. Quirke finds partially irreversible sorption isotherms for a model of nitrogen in a graphite slit-pore [173]. Peterson and Gubbins report metastable states for an atomic Lennard-Jones cylindrical vicinal phase with unstructured walls [169]. Similar results are obtained by Walton and Quirke [170] for an atomic Lennard-Jones fluid between structureless walls in slit-pore geometry and by Schoen et al. for model I [171].

However, the mere occurrence of metastable states in these calculations is not necessarily a serious problem in Wood's sense [58] because the true equilibrium state can be identified by computing ψ along the liquid- and gaslike branches of the sorption isotherms. Wherever these overlap the equilibrium state is the one having lower ψ. This logic is invoked in Walton and Quirke's careful and detailed study of capillary condensation [170].

Influence of Run Length and System Size. In Fig. 4.1 the plot of a sorption isotherm from grand canonical ensemble MC calculations for model I at a subcritical temperature gives clear evidence of metastability [171]. Following the sorption isotherm along the liquidlike branch by lowering the chemical potential leads to a spontaneous evaporation for $-\mu^* > 12.225$ while from the gaslike branch spontaneous condensation occurs for $-\mu^* < 12.150$.

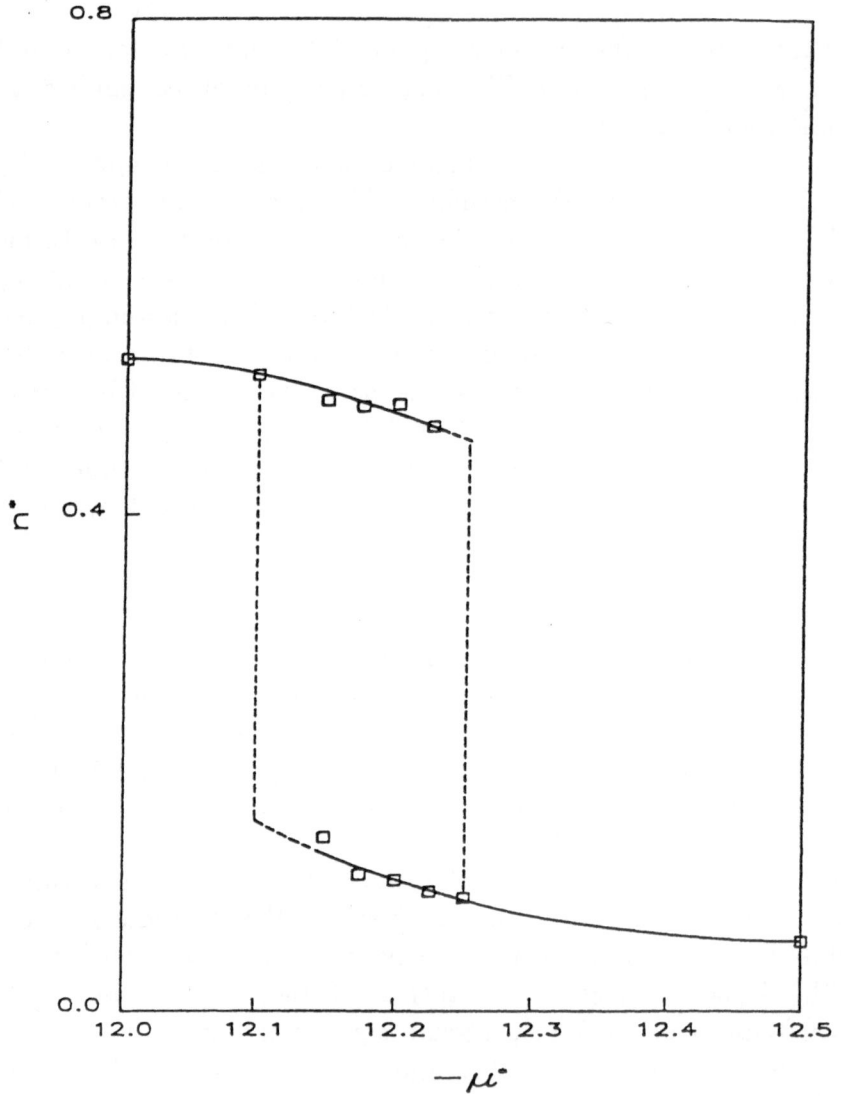

Fig. 4.1: Sorption isotherm at $h^* = 4.90$, $T^* = 1.00$ and $s^* = 7.9925$. Vertical dashed lines indicate range of chemical potentials within which metastability is observed; n^* denotes pore average density.

Thermodynamic states in the metastable loop $12.150 \leq -\mu^* \leq 12.225$ cannot easily be ascertained. A careful inspection of the Markov chain reveals that for the chosen system size (see Fig. 4.1) the vicinal phase oscillates between liquid- and gaslike branches of the sorption isotherm. This can be seen in Fig. 4.2 where the pore average density is plotted as a function of the number of MC steps. While the system "stays" predominantly on the gaslike branch it can undergo sudden transitions to the liquidlike branch. It remains in the higher density state for a large number of MC steps before another transition back to the lower density state occurs. The plot in Fig. 4.2 is representative of all states in the metastable loop. Thus, if MC runs are too short one might accidentally get trapped in states pertaining to one or the other branch of the sorption isotherm.

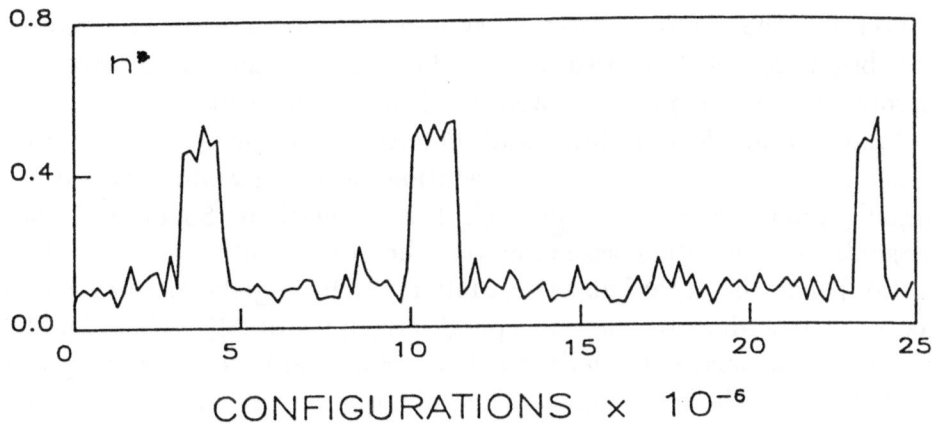

Fig. 4.2: Average pore density n^* as a function of the number of MC steps in the grand canonical ensemble at $h^* = 4.90$, $T^* = 1.00$, $s^* = 7.9925$ and $-\mu^* = 12.200$.

This notion suggests to interpret metastability not as hysteresis but as an ergodic problem in the sense of the discussion in the preceding section. A useful measure of the effort at exploring configuration space is the number of MC steps per particle. While this is of the order of 69000 in [171], Peterson and Gubbins allowed only for 15000 steps per particle in their runs [169]. For the plot in Fig. 4.2 this is comparable to the number of MC steps per particle the system spends on the liquidlike branch of the sorption isotherm and much less than the corresponding figure for the gaslike branch. The significance of Markov chain length with respect to metastability in capillary condensation is also noted by Walton and Quirke (see Fig. 5 in [170]) who are also very careful about the synonymous use of the terms "metastability" and "hysteresis".

From a comparison of density functional results for the single pore model and metastability in their grand canonical ensemble MC results Peterson and Gubbins infer that hysteresis occurs in their sorption isotherms at temperatures below T_c^{cap} [169]. However, this interpretation is based upon insufficiently long runs and, perhaps, too small systems employed in their simulations. Size of the MC cell is an important factor determining the extent of metastability [170,171]. In [171] it is demonstrated that enlarging the size of the MC cell in the (x,y)-plane (while holding h fixed) causes the size of the meta-stable loop in Fig. 4.1 to shrink. This comparts with Wood's remarks concerning the relation between system size and ergodicity cited at the beginning of the preceding section. His comment is certainly apropos of vicinal phases because of their confinement.

From the above it follows that Peterson and Gubbins' "hysteresis" must be regarded as a misinterpretation of metastability caused by ergodic problems involved in their MC simulations. Snook and van Megen's [167] sorption isotherms, on the other hand, (see preceding section) must be regarded as unadulterated by ergodic problems and "correct" according to the conceptual principles of MC. Thus, in MC every system must eventually reach thermodynamic equilibrium given long enough Markov chains. This philosophy prevents metastable states from having any real physical significance.

Metastability can be represented as a cusp catastrophy [171,174]. Fig. 4.3 displays a qualitative plot of n^* as a function of μ^* and h^* (which appears as a surface in three-dimensional space) for fixed T^* and s^*. The projection of this surface onto the (μ, h)-plane shows a bifurcation set, which corresponds to the metastable region in the vicinity of capillary coexistence. The bifurcation set represents meta-stable thermodynamic states of the vicinal phase, the vertex corresponding to the capillary critical point. The path EF above the critical point is reversible; n^* is a continuous function of μ^* along EF.

On the other hand, if the path traverses the bifurcation set, it is irreversible. This is illustrated by the following *Gedanken* experiment. Suppose the vicinal phase is initially liquidlike in thermodynamic state C (Fig. 4.3). As μ is gradually decreased, the state of the system moves along CD. At D the vicinal phase abruptly (i.e. "catastrophically") evaporates to form a gaslike phase in state A. This transition corresponds to the left bifurcation curve in the (μ^*, h^*)-plane. If μ^* is next increased the gaslike phase in state A

CUSP CATASTROPHE

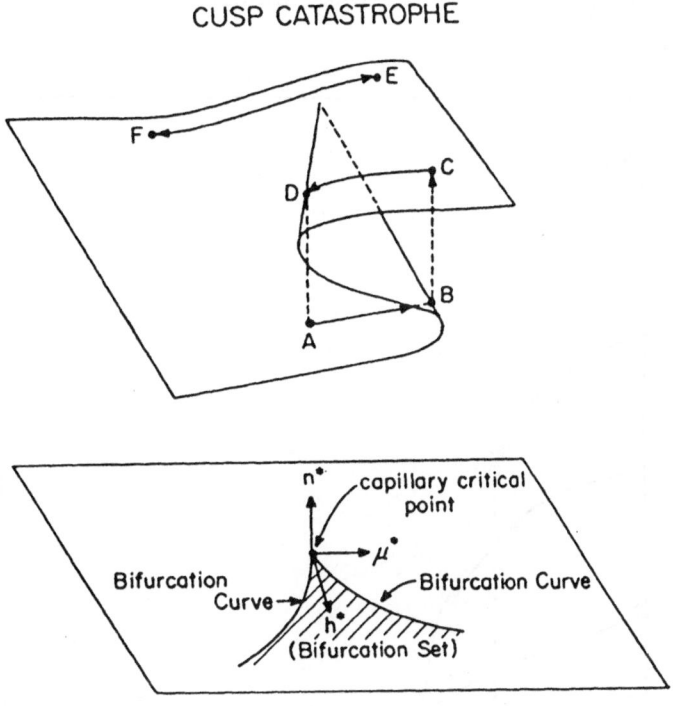

Fig. 4.3: The cusp catastrophe representing metastability in the slit-pore model.

does not instantaneously condense, but rather persists along AB until state B is reached whereupon it suddenly condenses to a liquidlike phase in state C. This transition is represented by the right bifurcation curve in the (μ^*, h^*)-plane. The isotherm plotted in Fig. 4.1 is a realization of the type of irreversible path just described. The loop CDABC in Fig. 4.3 is the counterpart of the loop in Fig. 4.1.

As shown in [171] the loops in subcritical sorption isotherms shrink with increasing system size. It can, therefore, be expected that in the infinite-system limit the bifurcation set shrinks to a *line* and all previously irreversible paths become reversible.

4.2.4 The Phase Diagram

If one corrects for metastability sorption isotherms change discontinuously according to a first order phase transition for temperatures below T_c^{cap} at capillary coexistence. An example is given in Fig. 4.4 where grand canonical ensemble MC results of [171] for model I are presented. The plots in Fig. 4.4 also indicate that the discontinuous

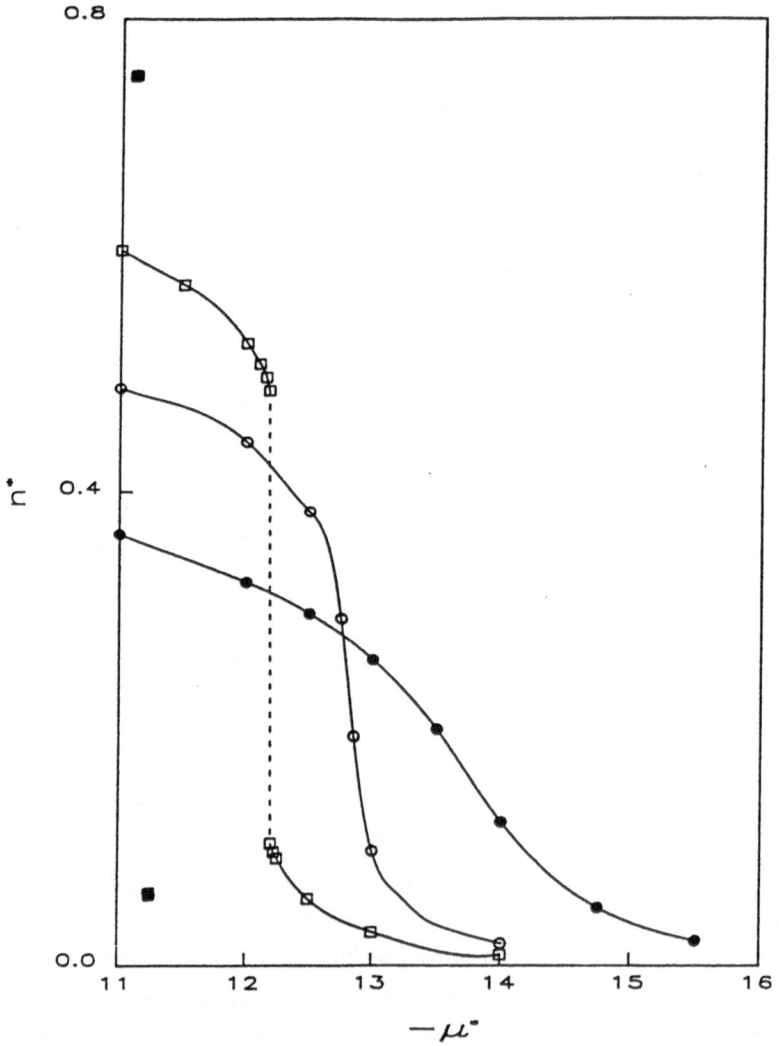

Fig. 4.4: Sorption isotherms at $T^* = 1.0$. **a:** $h^* = 2.15$ (●); **b:** $h^* = 3.05$ (◯); **c:** $h^* = 4.90$ (□). Dashed line indicates capillary coexistence (see text). Two points on the sorption isotherm of the bulk phase are shown (■).

change in the sorption isotherm disappears if h gets smaller. In addition the curves are shifted to lower chemical potentials as h increases. Between $3.05 < h^* < 4.90$ sorption isotherms become continuous functions of μ. This suggests existence of a capillary critical point (μ_c, h_c) in which the capillary coexistence curve terminates. The sorption isotherms at $h^* = 3.05, 2.15$ represent supercritical states in the spirit of the path EF in Fig. 4.4. These isotherms are perfectly reversible and show no evidence of metastability.

Also shown in Fig. 4.4 are two points pertaining to liquid and gaseous bulk phases close to the coexistence line. Apparently the liquid-gas transition occurs at higher μ in the bulk. The associated density change is also larger than at, say $h^* = 4.90$ in the corresponding vicinal phase. This qualitative observation complies with quantitative results obtained by Heffelfinger et al. [175] who show that confinement of fluid causes the width of the coexistence curve to shrink so that the gas- and liquidlike phases are much more similar in overall density compared with the bulk (see Fig. 6 in [175]). In addition $T_c^{cap} < T_c^{bulk}$. Similar observations are made experimentally [176].

4.2.5   Wetting

Wetting is a phenomenon closely related to capillary condensation. As described before capillary condensation can be viewed essentially as adsorption of a substance, say β on α where α represents the walls. Wetting is distinguished from capillary condensation because here β is not adsorbed on a uniform phase α but onto an interface between α and another phase γ [177]. α, β and γ can in principle represent any combination of solid and fluid (gas- and liquidlike) phases.

If the amount adsorbed diverges as temperature and chemical potential approach their values at β-γ-coexistence, β is said to completely wet the α-γ interface in the limit of α-β-γ coexistence. In general, one expects this to happen between the so-called wetting temperature T_w and the β-γ critical temperature [178]. At T_w partial wetting changes to complete wetting. For partial wetting the α-γ contact angle is non-zero and less than π, whereas for complete wetting this angle vanishes so that adsorption of β leads to formation of a layer separating α and γ. The wetting transition can be of first or second order. The order depends critically on the type of forces acting between the three phases [179].

Henderson and van Swol study wetting by MD for a solid-fluid-fluid system [179]. Their system consists of a square-well fluid adsorbed on planar hard walls. From their detailed analysis the authors deduce a first order wetting transition. They also compare their results with predictions from statistical-physical sum rules [27,131] and capillary wave theory [180]. In the sum rule approach both grand potential and local density are treated as functionals of $\mu - v(\mathbf{r})$ (see Sect. 3.5.2). A particularly important result obtained by Henderson and van Swol expresses the amount adsorbed in terms of an integral over a product of local density times transverse structure factor. Capillary wave theory enables one to reexpress the transverse structure factor yielding an expression which predicts the amount adsorbed to change logarithmically with small variations $\delta\mu$. The MD results confirm the logarithmic dependence (see Fig. 3b in [179]).

Wetting at an argon-solid carbon dioxide interface is investigated by Sokolowski and Fischer who use constant temperature MD [181]. Their results confirm thin-to-thick film transitions predicted by various density functional theories [182-184]. Investigations similar to Sokolowski and Fischer's are carried out by Finn and Monson [185] who perform MC simulations in an ensemble closely related to the isostress-isostrain ensemble introduced in Sect. 2.3.2. A first order wetting transition of the type reported by Henderson and van Swol [177] is also presented by Sikkenk et al. [186].

4.3 Shear Melting of Solid Vicinal Phases

The discussion of phase transitions in vicinal phases so far is exclusively concerned with liquid- and gaslike states. However, in Sect. 3.4.2. it is shown that vicinal phases may also exhibit solidlike characteristics depending on a number of important factors.

One of these factors is registry. This is apparent from plots displayed in panel (b) of Fig. 3.7 where the in-plane pair correlation function is shown for model I at $h^* = 2.15$ and two different values of $\alpha = 0.0, 0.5$. While only a broad layer of fluid is accomodated by the out-of-registry configuration ($\alpha = 0.5$) a solidlike vicinal phase is observed for the in-registry configuration ($\alpha = 0.0$). By virtue of its inherent structure the walls of model I exert a shear stress on the vicinal phase upon changes in registry (see eq. (3.4a)). This suggests that it might be possible to initiate transitions between liquid- and solidlike states by varying registry. For obvious reasons this shear strain induced "melting" is termed "shear melting" and is currently receiving a lot of attention by both theoreticians and experimentalists.

4.3.1 Introduction to Shear Melting Experiments

With a modification of the SFA described in Sect. 3.2.1 it is possible to investigate shear melting experimentally by essentially changing the registry between upper and lower walls as outlined in the previous section. As in the original (i.e. static) SFA two mica coated cylindrical silica lenses are positioned at right angles to one another.

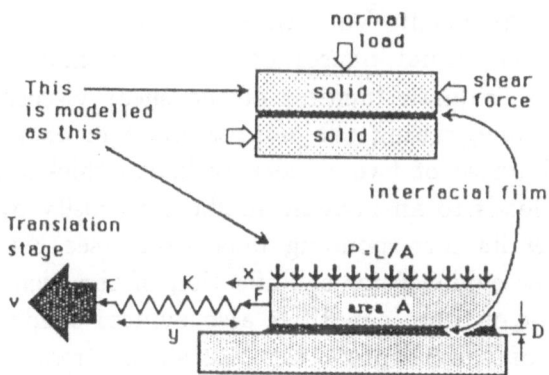

v	Velocity of translation stage
x	Total displacement of upper surface
v_x	Velocity of upper surface ($v_x = dx/dt$)
K	Spring stiffness
y	Spring length
F	Frictional force (spring tension = K Δy)
A	Contact area
L	Externally applied load
P	Externally applied pressure (P = L/A)
D	Thickness of film

Fig. 4.5: Schematic representation of the experimental setup to measure shear forces in vicinal phases (from [187]). For details see [188].

The modified setup allows the cylinders to be slid past one another at constant speed. Sliding takes place under fixed normal pressure (i.e. load) exerted externally on the crossed cylinders. Therefore, thickness of the vicinal phase may vary as sliding progresses. As in the static SFA separation between the surfaces can be determined with an accuracy of $\pm 1\overset{\circ}{\text{A}}$. Important new information from the dynamic SFA is the shear stress τ associated with the sliding process. Since only one surface is slid the shear force can be inferred from the measured displacement of the other one, x, to which a spring of known stiffness K is attached so that $\tau = K \cdot x$. A schematic representation of the experimental setup is depicted in Fig. 4.5. For a detailed description the interested reader is referred to [188].

Van Alsten and Granick use a similar piece of equipment in their investigation of liquid film properties under shear [189]. However, their method lacks fixed relative crystallographic orientation Θ as do earlier experiments by Israelachvili et al [190]; later modifications of the method allow control over Θ [191]. However, none of the dynamic SFA experiments can control the direction of sliding in the (x,y)-plane.

From their experiments Van Alsten and Granick deduce the shear viscosity for thin vicinal phases of non-polar molecules [189]. The results reveal a dramatic increase of the shear viscosity in very thin films. For example, raising the normal pressure exerted on a hexadecane vicinal phase of two molecular layers thickness by roughly a factor of two leads to an increase in shear viscosity by four orders of magnitude over its corresponding bulk value (see Figs. 2,3 in [189]). This is taken as evidence for solidification of the vicinal phase.

The behavior of other quantities also evinces solidification. Fig. 4.6 displays plots of various properties like shear stress, lateral displacement of the sliding surface and thickness of the vicinal phase as functions of time for a typical shear melting experiment. Starting from a solidlike vicinal phase a linear increase of the shear stress is found at first (upper panel, Fig. 4.6) corresponding to the response of an elastic solid to shear strain. After a certain yield value is reached τ drops rapidly suggesting a solid-liquid phase transition. This melting process is accompanied by a sudden increase of film thickness (third panel, Fig.4.6).

The sliding speed v is critical for details of shear melting. If v exceeds a certain value v_c, solidlike phases do not form again once they are molten. This can be seen from Fig. 4.6 where the film

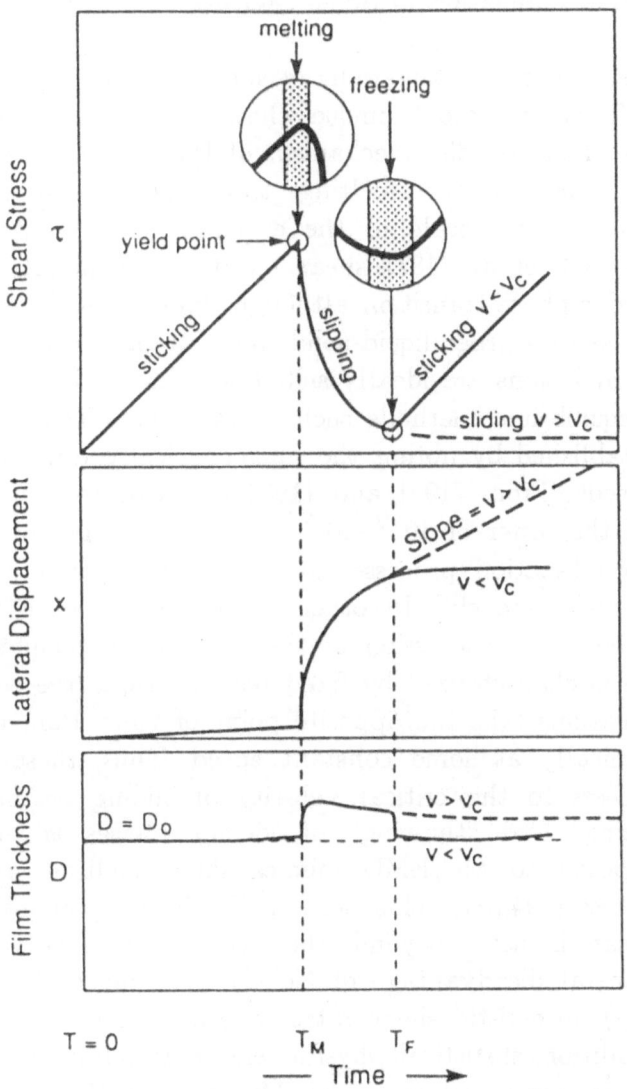

Fig. 4.6: Changes of various vicinal phase properties during shear melting (after [187]).

thickness does not change back to its smaller value in the solidlike phase after melting. On the contrary, if $v < v_c$ periodic "freezing" and "thawing" is observed. The critical speed depends on a number of parameters like relaxation times and the possibility to form solidlike structures commensurate with the walls. The "ease" with which solidlike vicinal phases can form will determine v_c.

4.3.2   Computer Simulation of Shear Melting

Although experimental investigations of shear melting reveal a lot of remarkable information about unique characteristics of thin vicinal films, they give none on the mechanism of the melting process. For example, the order of the melting transition is unknown and inaccessible on the time scale of the experiment [187]. As pointed out by Israelachvili et al. [187] shear melting is generally believed to be a first order phase transition although data do not preclude the possibility of a second order liquid-solid phase transition (see Fig. 4.6).

Computer simulations are ideally suited to address this question. A link between equilibrium methods such as MC or EMD and the experiment is established by noting the microscopically low experimental sliding speeds. From [192] and [193] one can readily compute these to be of the order of 10^{-9}–$10^{-7}\,\overset{\circ}{A}\,ps^{-1}$. Thus, one may assume that molecular relaxation processes are much faster than the speed at which the walls are slid. In other words, in MC or EMD shear melting is viewed as a succession of thermodynamic equilibrium states. Each state is characterized by fixed registry, i.e. α (see eq. (3.4a)).

NEMD simulations take the opposite point of view. Here the walls are slid dynamically at some constant speed. Thus, these methods have direct access to the critical velocity of sliding beyond which periodic "freezing" and "thawing" of vicinal phases is prohibited. NEMD as opposed to MC/EMD mimics shear melting as a rate controlled process [194,195]. This permits NEMD to gain insight into details of shear melting beyond the regime of linear response. However, a general disadvantage of NEMD is its unavoidable use of (experimentally) unrealistic shear rates of approximately 1–$10\,\overset{\circ}{A}\,ps^{-1}$ [193,196]. In addition, statistical-physical ensembles employed are less rigorously defined compared with, say MC so that the reliability of NEMD results needs to be established by comparison with experiments or other independent simulation data.

Shear Melting in the Grand Canonical Ensemble. The first computer simulation of shear melting was carried out by Schoen et al. [197] by means of grand canonical ensemble MC. Based upon results discussed in Sect. 3.4.2 shear melting is investigated for model I at three (fixed) interwall separations $h^{*} = 2.20$, 3.10 and 4.90. According to plots in panels (b), (c) and (e) of Figs. 3.5 and 3.7 two solidlike layers form at $h^{*} = 2.20$ and $\alpha = 0.5$ (each layer comprises about 50 atoms; $s^{*} = 7.9925$)

while three and five solidlike layers are accommodated at $h^* = 3.10$ and 4.90 respectively, at $\alpha = 0.0$.

By analogy with eq. (3.12) shear stress is given by

$$\langle \tau_{zx}^{(2)} \rangle = -\langle F_x^{(2)} \rangle = s^{-2} \left\langle \sum_{i=1}^{N} \sum_{j=1}^{N_s} \frac{du(r_{ij}^{(2)})}{dr_{ij}^{(2)}} \frac{x_{ij}^{(2)}}{r_{ij}^{(2)}} \right\rangle \qquad (4.2)$$

By symmetry there is an analogous expression for $\langle \tau_{zx}^{(1)} \rangle$. Physically, $\tau_{zx}^{(k)}$ is the x component of the force exerted by the vicinal phase on wall k. Again by symmetry

$$\langle \tau_{zx}^{(2)} \rangle = -\langle \tau_{zx}^{(1)} \rangle \qquad (4.3)$$

so that one can average over both walls with respect to the difference in sign in eq. (4.3). This average will be denoted $\langle \tau_{zx} \rangle$

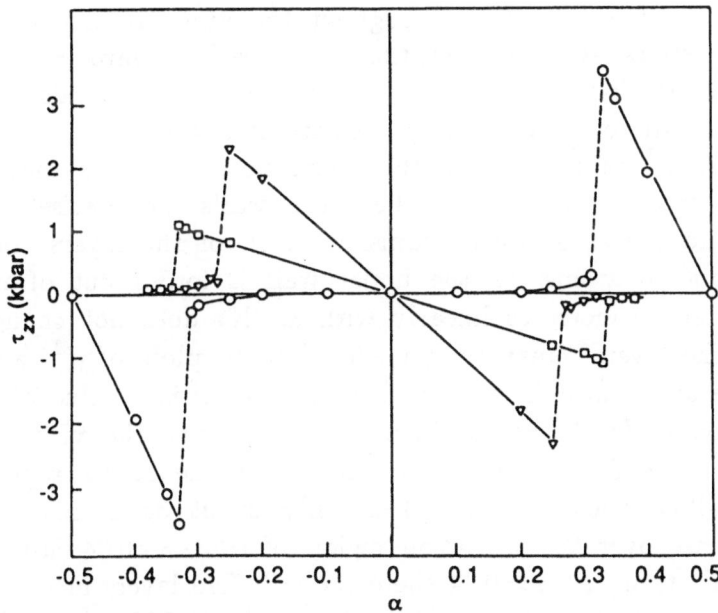

Fig. 4.7: Shear stress τ_{zx} as a function of strain $\sim \alpha$ for walls separated by $h^* = 2.20$ (○), 3.10 (△) and 4.90 (□); $-\mu^* = -9.26$; $T^* = 1.0$.

henceforth. Because $\langle \tau_{zx} \rangle$ must be periodic in the registry α, with a period of unity (see eq. (3.4a)) plots in Fig. 4.7 are shown only in the range $-0.5 \leq \alpha \leq 0.5$ [197]. By virtue of the symmetry of the walls, $\langle \tau_{zx} \rangle$ is antisymmetric in α. The shapes of shear stress curves are similar but the curve for $h^* = 2.20$ differs in phase by $\Delta\alpha = 0.5$ from the other two.

For $h^* = 2.20$ the linear portion of the shear stress curve is centered on $\alpha = 0.5$ where the walls are fully out of registry and $\langle \tau_{zx} \rangle$ vanishes exactly. As the walls are sheared so that α decreases from 0.5, shear stress rises approximately linearly with α down to $\alpha \approx 0.33$, tending to restore the walls to $\alpha = 0.5$. The average number of atoms in the vicinal phase remains essentially constant; plots of $\rho^{(1)}$ and $g^{(2)}$ indicate two distorted fcc (100) layers sharply localized in the z-direction over this range of α. Shear stress attains a maximum value at about $\alpha = 0.33$. This is the critical stress required to initiate sliding. If the walls are strained beyond this point, shear stress decreases abruptly. Simultaneously $\langle N \rangle$ decreases by about 44 as α goes from 0.33 to 0.31. Almost an entire layer of atoms suddenly exits the vicinal phase over this range of α. Plots of $\rho^{(1)}$ and $g^{(2)}$ now indicate the presence of a broad single fluidlike layer (cf. panel (b) in Figs. 3.5, 3.7). As α decreases further, $\langle N \rangle$ gradually decreases to about 50 at $\alpha = 0.0$, but throughout the range from 0.33 to 0.00 the shear stress acts to push the walls back toward $\alpha = 0.5$. The point at $\alpha = 0.0$ is metastable; any slight displacement tends to return the upper wall to the equilibrium registry at $\alpha = 0.5$.

For both $h^* = 3.10$ and 4.90 the linear portion of the shear stress curve is centered on $\alpha = 0.0$, where the walls are precisely out of registry and the shear stress vanishes. Each of the layers comprises again about 50 atoms. As the upper wall is forced out of registry, the shear stress increases linearly with α; $\langle N \rangle$ does not change and the solidlike layers remain intact according to plots of $\rho^{(1)}$ and $g^{(2)}$. When the upper wall is strained beyond the critical value of α, $\langle N \rangle$ drops sharply. For $h^* = 3.10$ about 41 atoms leave the vicinal phase over the range $0.25 < \alpha < 0.27$; two broad fluidlike layers replace the three solidlike ones. For $h^* = 4.90$, only about 35 atoms exit the vicinal phase over the transition region of $0.33 < \alpha < 0.34$ just beyond the critical strain. At $\alpha = 0.34$ there are still five layers in the vicinal phase. The contact layer remains a distorted fcc (100) plane; the first inner layer retains a trace of fcc character; the innermost layer, which is sparsely populated, is strictly fluidic.

The stress curves have the following features. Both the slope of the linear region (the "force constant") and the critical stress decrease as the number of solidlike layers increases. However, the range of α over which solidlike behavior persists increases with the number of layers; that is, the critical strain increases with the number of layers. The critical strain per layer remains constant at about 0.08. With increasing h, the shear stress curve becomes flatter. At sufficiently large h, the pore should become stable at all registries α.

The decrease of critical stress with increasing number of layers is also observed experimentally [190] and can be rationalized as follows. As the number of layers increases, the solidlike character of each layer decreases toward the center of the vicinal phase and it takes less force to break down the less ordered structure of the inner layer.

Shear Melting in the Isostress-Isostrain Ensemble. Shear melting in the grand canonical ensemble occurs at fixed h so that the normal pressure on the walls varies with α. In the corresponding experiments, however, the load is fixed but h may vary. To better account for the experimental conditions the isostress-isostrain ensemble is employed here. The isostress-isostrain ensemble is introduced in Sect. 2.3.2 where a modified version of the Metropolis algorithm is described which permits the generation of a numerical representation of Markov chains in this ensemble. It is noted that the isostress-isostrain ensemble is closely related to the isothermal-isobaric ensemble for bulk fluids [198,199]. However, while in the latter the bulk phase can only be contracted or expanded isotropically, the isostress-isostrain ensemble allows for anisotropic compression and expansion as well as for changes of vicinal phase shape (see Sect. 2.3.2).

Structure and Shear Stress. In the isostress-isostrain ensemble the thermodynamic state of the vicinal phase is completely determined by specifying N, T, normal stress τ_3 (denoted τ_{zx} in Sect. 4.3.2) and strain components σ'_3 (see Sec. 2.3.2). That is, vicinal phase properties depend on α via strain component σ_4 (see eq. (2.22d)). h (i.e. σ_3) is no longer fixed as in grand canonical ensemble MC calculations but may vary now subject to the conjugate (fixed) stress τ_3.

In this section vicinal phase structure is exclusively investigated in terms of the in-plane pair correlation function $g^{(2)}$ defined in eq. (3.20). As in the previous section all the results pertain to model I.

Plots of $g^{(2)}$ in Fig. 4.8 indicate the complexity of shear melting in the isostress-isostrain ensemble for a monolayer vicinal phase [64]. A solidlike vicinal monolayer forms at $\alpha = 0.0$ ($\tau_3 = 0.0$, $T^* = 1.0$, $N = 50$, $s^* = 7.9925$) indicated by the strongly peaked $g^{(2)}$. Structural characteristics are the second neighbor shell peak at $\rho_{12}^* \approx 1.6$ and the overlapping peaks in the range $\rho_{12}^* = 2.3 - 2.5$, which are incompletely

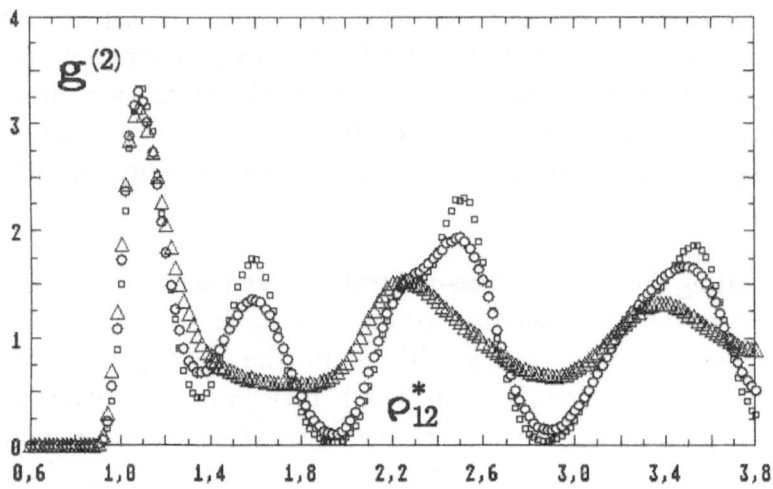

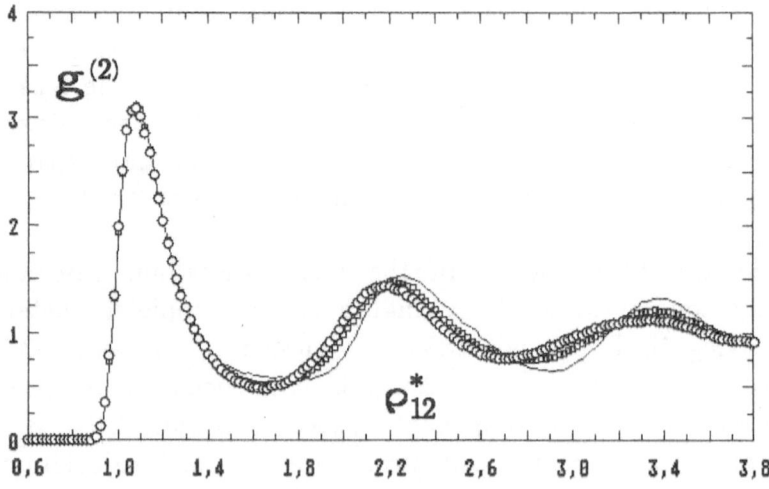

Fig. 4.8. In-plane pair correlation function for monolayer vicinal phases. **a:** $\alpha = 0.0$ ($\triangle$); $\alpha = 0.15$ ($\square$); $\alpha = 0.30$ ($\circ$). **b:** $\alpha = 0.30$ (——); $\alpha = 0.35$ ($\square$); $\alpha = 0.50$ ($\circ$).

resolved due to thermal motion. Again peak positions are identical with those associated with the (100) plane of the fcc lattice. At $\alpha = 0.15$ the structure is much less pronounced at larger ρ_{12}. Nevertheless, the second neighbor shell peak is still visible and so is a shoulder at $\rho_{12}^* \approx 2.3$, which is the remnant of the formerly unresolved double peak. Yet, the vicinal phase is still solidlike, as the sustained long range order indicates. At $\alpha = 0.30$, however, the fcc features are no longer present in $g^{(2)}$. Most noticeable is the absence of a peak at $\rho_{12}^* \approx 1.6$ and the shift to smaller ρ_{12} for all peaks but the first. Although much less pronounced, long range order is discernible in $g^{(2)}$ at this α as well. The dependence of $g^{(2)}$ on α suggests that shear melting occurs over the range $0.175 < \alpha < 0.35$.

It is also instructive to investigate the dependence of various stress tensor components on laterally applied strain σ_4. Following the notation developed in Sect. 2.3.2 (see also [64]), microscopic expressions for the stress tensor components can be derived from the statistical-physical expression

$$\langle \tau_\alpha \rangle = -\beta \left(\partial \ln Q_{N\tau_3 T} / \partial (V^0 \sigma_\alpha) \right)_{T, N, \sigma'_{3,\alpha}} \tag{4.4}$$

where $Q_{N\tau_3 T}$ is the isostress-isostrain partition function; $\sigma'_{3,\alpha}$ denotes differentiation with respect to fixed strain tensor components except for σ_α and σ_3 (see eqs. (2.22)). Carrying out the somewhat lengthy but straightforward algebraic manipulations one finally arrives at [64]

$$\langle \tau_\alpha \rangle = \langle \tau_{\alpha,pp} \rangle + \langle \tau_{\alpha,pw} \rangle \tag{4.5}$$

where

$$\langle \tau_{\alpha,pp} \rangle = -N(\beta s^2)^{-1} \langle h^{-1} \rangle + (s^2 \langle h \rangle)^{-1} \left\langle \sum_{i=1}^{N-1} \sum_{j=i+1}^{N} \frac{du(r_{ij})}{dr_{ij}} \frac{\alpha_{ij}^2}{r_{ij}} \right\rangle \tag{4.6a}$$

and

$$\langle \tau_{\alpha,pw} \rangle = (s \langle h \rangle)^{-1} \left\langle \sum_{k=1}^{2} \sum_{i=1}^{N} \sum_{j=1}^{N_s} \frac{du(r_{ij}^{(k)})}{dr_{ij}^{(k)}} \frac{\alpha_{ij}^{(k)2}}{r_{ij}^{(k)}} \right\rangle \tag{4.6b}$$

80

and $\alpha = x(1)$, $y(2)$. From eqs. (3.12) and (3.13) one also finds

$$2s^2 \, \bar{\tau}_3 = \left\langle F_z^{(1)} \right\rangle - \left\langle F_z^{(2)} \right\rangle \tag{4.7}$$

which provides a consistency check for the MC simulation because $\bar{\tau}_3$ has to equal the input value of τ_3. The plot of $\bar{\tau}_3$ versus α in Fig. 4.9 indicates that the MC runs presented in this section satisfy

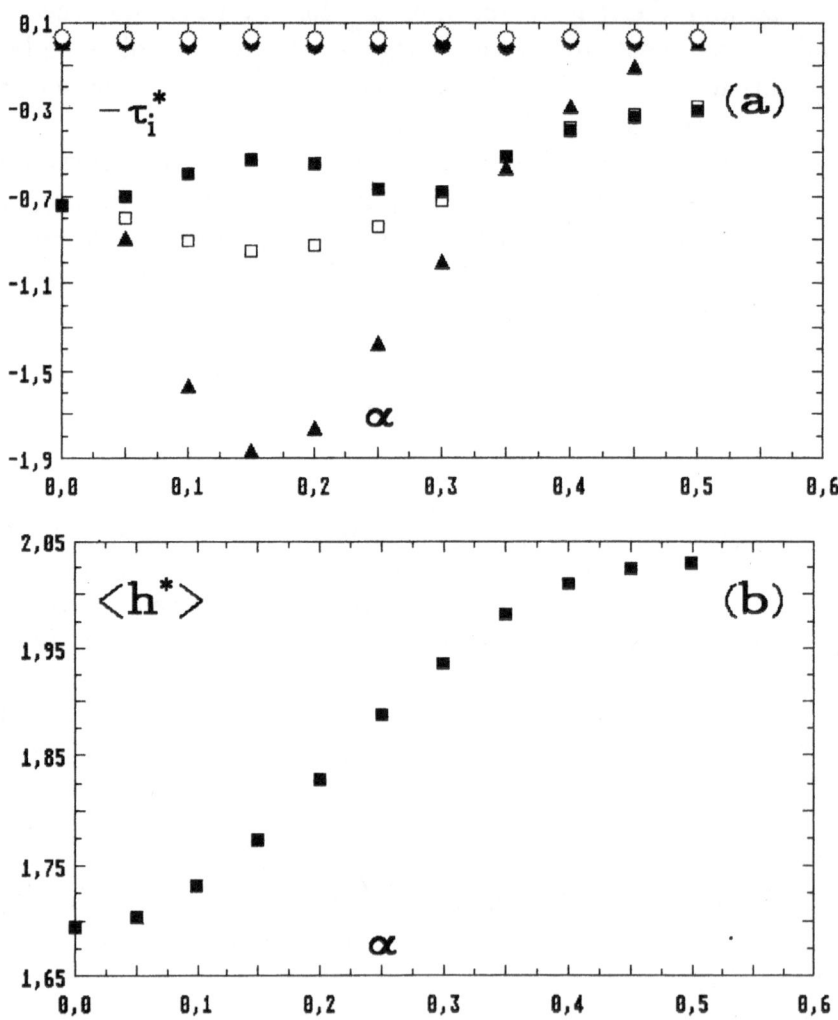

Fig. 4.9: **a**: Negative stress tensor components $-\tau_i^*$ as functions of α at $\tau_3^* = 0.0$, $T^* = 1.0$. $-\langle \tau_1 \rangle$ (■); $-\langle \tau_2 \rangle$ (□); $-\tau_3$ (○); $-\langle \tau_4 \rangle$ (▲); $-\langle \tau_5 \rangle$ (●). **b**: Average wall separation $\langle h^* \rangle$ as a function of α at $\tau_3^* = 0.0$, $T^* = 1.0$.

the relation $\tau_3 = \bar{\tau}_3$. Similarly, from eqs (4.2) and (4.3) it follows by analogy that

$$2s^2\langle\tau_4\rangle = \langle F_x^{(1)}\rangle - \langle F_x^{(2)}\rangle \tag{4.8a}$$

$$2s^2\langle\tau_5\rangle = \langle F_y^{(1)}\rangle - \langle F_y^{(2)}\rangle \tag{4.8b}$$

However, in the present case $\langle\tau_5\rangle = 0.0$ regardless of α for symmetry reasons because shear strain is only applied along the x-direction ($\alpha' = 0.0$, see eq. (3.4b)). This is confirmed by plots in Fig. 4.9.

Except for $\bar{\tau}_3$ and $\langle\tau_5\rangle$ the stress tensor components exhibit an interesting dependence on shear strain σ_4. As α increases from 0.0, $\langle\tau_4\rangle$ increases approximately linearly until $\alpha \approx 0.1$ (cf. Fig. 4.7). This is the expected response of an elastic solid to small strain. As α increases beyond 0.10, the *rate of increase* of $\langle\tau_4\rangle$ decreases markedly until α reaches a maximum at $\alpha \approx 0.18$, after which it decreases, eventually vanishing at $\alpha = 0.50$, where the walls are completely out of registry. Note the striking difference between $\langle\tau_4\rangle$ in Fig. 4.9 and the corresponding curves in Fig. 4.7. Apparently, expansion of the vicinal phase in the isostress-isostrain ensemble provides a much more efficient mechanism to release shear stress compared with drainage in the grand canonical ensemble. Drainage can occur only by removing an entire layer from the vicinal phase, whereas a constant number of particles can reorganize themselves spatially by only a slight increase in the wall separation (see Fig. 4.9b). Thus, in the isostress-isostrain ensemble successive relative wall alignments cause structurally slightly differing vicinal phases so that properties of these phases depend continuously on α (i.e. on σ_4); by contrast, removal of an entire layer changes the structure of the vicinal phase rather abruptly.

The transverse compressional stresses $\langle\tau_1\rangle$ and $\langle\tau_2\rangle$ are equal in the unstrained solidlike isotropic monolayer ($\alpha = 0.0$). Shearing causes $\langle\tau_1\rangle$ to decrease with increasing α until a (relative) minimum is reached, which is very near the maximum in $\langle\tau_4\rangle$ at $\alpha \approx 0.18$. On the other hand, $\langle\tau_1\rangle$ increases with α to a (relative) maximum that roughly coincides with the minimum in $\langle\tau_2\rangle$. Beyond $\alpha \approx 0.30$, where it is surmised the monolayer is molten, and therefore isotropic, $\langle\tau_1\rangle \simeq \langle\tau_2\rangle$.

It is again instructive to compare $g^{(2)}$ for three different strains in this range $0.30 \leq \alpha \leq 0.50$ (see Fig. 4.8). As before one notes the pronounced long range order at $\alpha = 0.30$, which is the greatest strain at which $\langle\tau_1\rangle \neq \langle\tau_2\rangle$. If the strain is increased further, these stress components eventually become equal and the corresponding $g^{(2)}$ reflects a much more disordered structure: peaks are shifted to smaller interatomic separations and are much more damped.

Comparing $g^{(2)}$ at $\alpha = 0.35$ with that at $\alpha = 0.50$, where the walls are completely out of registry, one is surprised that the structure of the vicinal phase seems to have changed little, since at the same time $\tau_\parallel = (\langle\tau_1\rangle + \langle\tau_2\rangle)/2$ changes by a factor of 1.73. Throughout the disordered regime $0.35 \leq \alpha \leq 0.50$ the vicinal phase exhibits considerable shear stress indicating its non-Newtonian character. These results parallel qualitatively the curves shown in Fig. 4.6 after the solidlike vicinal phases are molten.

Based upon the observation of continuity of stress tensor components as α changes it is already tempting to speculate that shear melting may be describable as a second order phase transition in the isostress-isostrain ensemble. However, continuity of stress tensor components in Fig. 4.9 does not suffice to determine unambiguously the order of the phase transition. Further evidence for a second order phase transition is provided by results discussed subsequently and in Sect. 5.3.

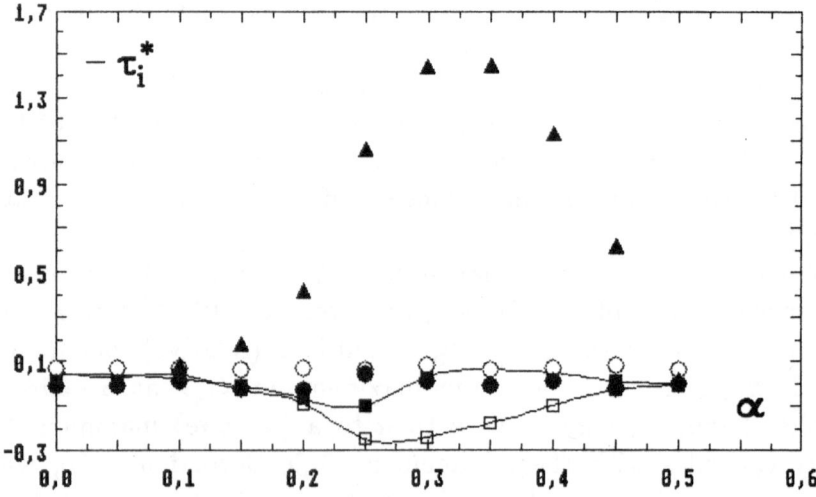

Fig. 4.10: As Fig. 4.9 but for a vicinal phase accomodating two layers. The full lines represent a cubic spline fit to the MC data to guide the eye.

The Effect of Vicinal Phase Dimensionality and Thermodynamic Conditions. None of the features discussed above depends qualitatively on T or τ_3 as shown in [64]. Increasing τ_3 generally stabilizes the solidlike vicinal phase. This is in accord with the experimentally observed dependence of shear viscosity on load [189] (see also Sect. 4.3.1). An increase in temperature, on the other hand, destabilizes solidlike vicinal phases. It is also demonstrated in [64] that the

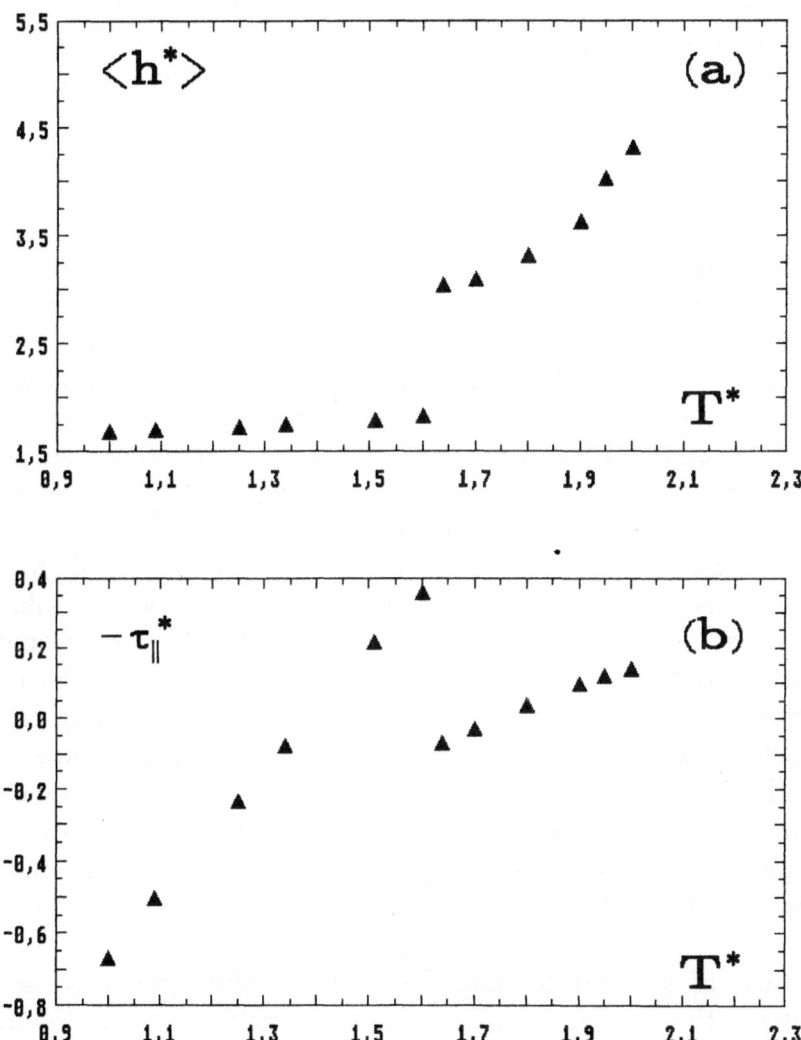

Fig. 4.11. a: Average wall separation $\langle h^* \rangle$ as a function of T^* at $\alpha = 0.0$, $\tau_3^* = -0.24$. **b:** $-\tau_{\parallel}^* = -(\langle \tau_1^* \rangle + \langle \tau_2^* \rangle)/2$ as a function of T^* at $\alpha = 0.0$, $\tau^* = -0.24$.

results presented in this section do not depend on enlarging the size
of the vicinal phase in the (x,y)-plane indicating that even a "small"
system of 50 vicinal atoms pertains already to the thermodynamic limit.

In addition, structural ($g^{(2)}$) and mechanical ($\{\tau_\alpha\}$) characteristics
of shear melting remain qualitatively unaffected if one abandons the
monolayer vicinal phase in favor of a two-layer vicinal phase. In
Fig. 4.10 stress tensor components are plotted as functions of α for
such a system. As mentioned before (panel (b) in Fig. 3.7) a solidlike
vicinal phase forms isotropically with the walls completely out of
registry ($\alpha = 0.50$). It is important to note that the apparent continuity
of the phase transition is not a unique feature of monolayer vicinal
phases.

Although the continuous nature of shear melting appears to be
qualitatively unaffected by changes of details of the system or of
thermodynamic conditions, the character of the transition alters
markedly if melting is effected by increasing temperature under constant load with no applied strain. This can be seen from Fig. 4.11
where mean wall separation $\langle h^* \rangle$ and $-\tau_\parallel^*$ are plotted as functions
of T^* at $\tau_3^* = -0.24$ and $\alpha = 0.0$ (cf. Fig. 4.9). Both plots comprise
two branches. In the low temperature regime up to $T^* = 1.60$, $\langle h^* \rangle$
increases linearly with a very small slope. $g^{(2)}$ for this monolayer
vicinal phase exhibits solidlike character throughout this regime. Since
$\langle h^* \rangle$ changes very little with temperature up to $T^* = 1.60$, $\tau_\parallel$ depends strongly on T and increases linearly, as expected for an
elastic solid. Somewhere in the vicinity of $1.60 < T^* < 1.64$ the solidlike
vicinal phase melts abruptly. $g^{(2)}$ establishes the fluidlike nature of
the new phase at $T^* = 1.64$. Within the finite resolution of the plots,
it appears that at constant load temperature-driven melting is discontinuous (i.e. a first order phase transition). For the high temperature
branch ($T^* > 1.64$) $\langle h^* \rangle$ increases monotonically and highly nonlinearly.
Simultaneously $\tau_\parallel$ decreases nearly linearly, but with a distinctly
smaller slope than the "low" temperature branch ($T \leq 1.60$), which is
not surprising for a dense liquidlike phase.

Localization of a Shear Melting Point. The results presented so far
demonstrate the continuous nature of shear melting in the isostress-
isostrain ensemble but do not allow one to locate a melting point in
terms of, say a critical strain. In fact, it is quite uncertain whether
a unique melting point exists due to the continuous and gradual
change of structural and mechanical properties studied so far.

Detection and localization of a solid-liquid transition point would shed more light on the mechanism of shear melting and, perhaps, allow one to determine the order of the phase transition. The latter is a particularly interesting aspect because it is believed [38] that melting in three dimensional systems can occur only as a discontinuous, first order phase transition accompanied by release of latent heat. From the results presented thus far this notion can be confirmed in part with the proviso that the melting process must be temperature-driven (see Fig. 4.11). In truly two-dimensional systems, on the other hand, several researchers have speculated on the existence of second order liquid-solid phase transitions [200].

A second order phase transition is rigorously characterized by a number of features. Most important, it occurs without release of latent heat, that is internal energy or enthalpy remain continuous functions of the external field (e.g. temperature, strain etc.) driving the melting process. Similarly, as observed in Fig. 4.9b, density is a continuously varying function. At the transition point all these quantities may change with infinite rate (see, for example Fig. 1.1b in [18]). Because of this it is possible to locate the transition point through divergences of heat capacity or compressibility at the transition point which are defined as derivatives of internal energy (enthalpy) or density. In statistical physics it is generally possible to express quantities such as heat capacity or compressibility in terms of fluctuations of internal energy (enthalpy) or density. Thus, from a statistical point of view it is the extent of fluctuations which diverges during a phase transition.

Definition of Fluctuation-Related Quantities. Quantities of prime interest in this section are isostress heat capacity

$$c_{\tau_3} \equiv \left(\frac{\partial H}{\partial T}\right)_{\tau_3} \tag{4.9}$$

isothermal compressibility

$$\varkappa_T \equiv -\left(\frac{\partial \langle \sigma_3 \rangle}{\partial \tau_3}\right)_T \tag{4.10}$$

and isostress expansivity

$$\gamma_{\tau_3} \equiv \left(\frac{\partial \langle \sigma_3 \rangle}{\partial T} \right)_{\tau_3} \tag{4.11}$$

where strain tensor component σ_3 is given in eq. (2.22c) and the iso-stress-isostrain enthalpy is defined as

$$H \equiv U - \tau_3 V^0 \langle \sigma_3 \rangle \tag{4.12}$$

In eq. (4.12) U denotes internal energy and the volume of the un-strained system is given by

$$V^0 = s^2 h^0 = s^2 \langle h \rangle \tag{4.13}$$

Expressing U and $\langle \sigma_3 \rangle$ in terms of the isostress-isostrain ensemble probability density in the classical limit (see [201]) allows one to reexpress eqs. (4.9)-(4.11) as

$$c_{\tau_3} = \frac{3}{2} N k_B + \frac{1}{N k_B} \left[\langle H'^2 \rangle - \langle H' \rangle^2 \right] \quad , \tag{4.14}$$

$$\varkappa_T = \beta s^2 \langle h \rangle^{-1} \left(\langle h \rangle^2 - \langle h^2 \rangle \right) \tag{4.15}$$

and

$$\gamma_{\tau_3} = \frac{\beta}{T \langle h \rangle} \left(\langle H'h \rangle - \langle H' \rangle \langle h \rangle \right) \tag{4.16}$$

where

$$H' = \Phi(\mathbf{r}^N;\mathbf{\sigma}) - \tau_3 s^2(h - \langle h \rangle) \tag{4.17}$$

is the configurational contribution to H. Splitting H into kinetic and configurational contributions relies on a conservative system hamiltonian. Note, that the kinetic contribution to c_{τ_3} (first term on the r.h.s. of eq. (4.14)) assumes an atomic system. From eq. (4.17) $\langle H' \rangle = \langle U \rangle$ because the second term on the r.h.s. of eq. (4.17) vanishes when averaged but $\langle H'^2 \rangle \neq \langle U^2 \rangle$.

Eqs. (4.14)-(4.17) can be employed to compute c_{τ_3}, x_T and γ_{τ_3} in the isostress-isostrain ensemble. Invoking purely thermodynamic arguments it may also be shown [201] that the three quantities are related via

$$c_{\tau_3} - c_{\sigma_3} = s^2 \langle h \rangle T \, \frac{\gamma_{\tau_3}^2}{x_T} \tag{4.18}$$

where

$$c_{\sigma_3} \equiv \left(\frac{\partial U}{\partial T} \right)_{\sigma_3} \tag{4.19}$$

is the isostrain heat capacity. Note, that c_{σ_3} cannot be calculated directly in the isostress-isostrain ensemble because strain tensor component σ_3 is not fixed but fluctuates around $\langle \sigma_3 \rangle = 0$. However, it may readily be shown that

$$c_{\sigma_3} = \frac{3}{2} N k_B + \frac{1}{k_B T^2} \left[\langle \Phi^2(\mathbf{r}^N;\mathbf{\sigma}) \rangle - \langle \Phi(\mathbf{r}^N;\mathbf{\sigma}) \rangle^2 \right] \tag{4.20}$$

which is obtainable from fluctuations of Φ in the canonical ensemble. Eqs (4.18) and (4.20) provide a consistency check for the MC results.

Strain Dependence. It is shown in Sect. 4.3.2 that monolayer vicinal phases (model I) solidify at $\alpha = 0.0$ ($T^* = 1.0$, $\tau_3^* = 0.0$); at $\alpha = 0.50$ the vicinal phase exhibits characteristics of a dense Lennard-Jones liquid, that is it lacks long range spatial order. From a study of strain dependence of $g^{(2)}$ and various stress tensor components it is concluded that melting occurs somewhere between these extremes. This becomes manifest, for example, in a steady and continuous increase of film thickness proportional to $\langle h \rangle$ (see Fig. 4.9b) or a similar increase of $\langle H' \rangle$ (eq. (4.17)) in Fig. 4.12. $\langle H' \rangle$ remains a continuous function of applied strain α throughout the entire range, that is shear melting occurs without release of latent heat, which is one of the characteristics of a second order phase transition [38].

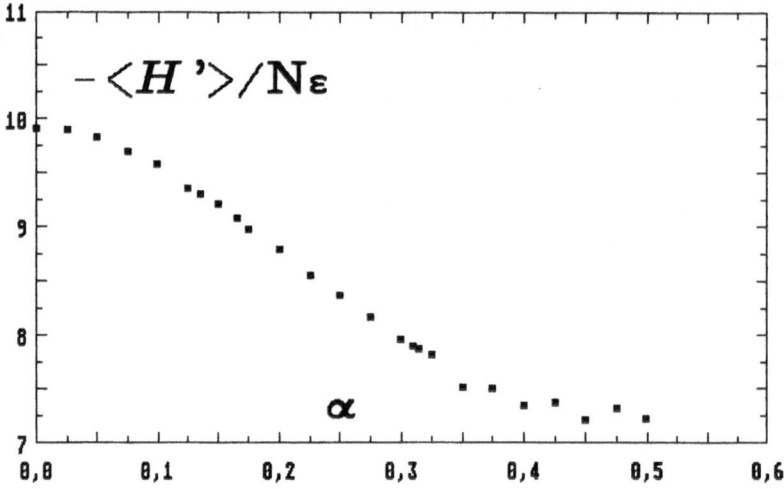

Fig. 4.12. Configurational contribution to the isostress-isostrain enthalpy $\langle H' \rangle$ per particle as a function of applied strain $\sim\!\alpha$.

Fluctuations of h and H' around their respective average values depend differently on α as can be seen from Fig. 4.13 where $\varkappa_T$, $\gamma_{\tau 3}$ and $c_{\tau 3}$ are plotted as functions of applied strain. Although details in the strain dependence differ between the three quantities [201], some features are common to them. Most important, they assume maxima in the same range of α indicating divergence in the vicinity of the shear melting point in accordance with considerations in this section.

To test consistency of these results $c_{\sigma 3}$ is employed as outlined at the end of the preceding section. As pointed out there $c_{\sigma 3}$ is identical with the usual isochoric heat capacity c_V which can be obtained from canonical ensemble MC calculations with $V = s^2 \langle h \rangle = f(\alpha)$ as thermodynamic state parameter instead of τ_3, $\langle h \rangle$ being the average separation between the walls obtained from a previous isostress-iso-

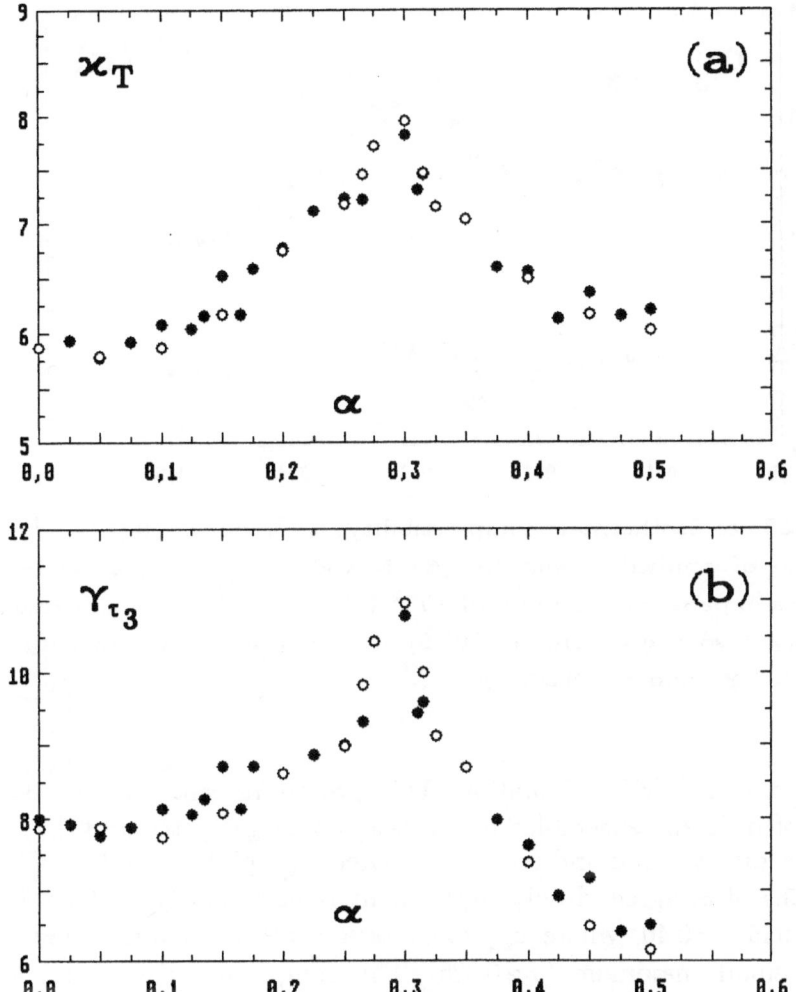

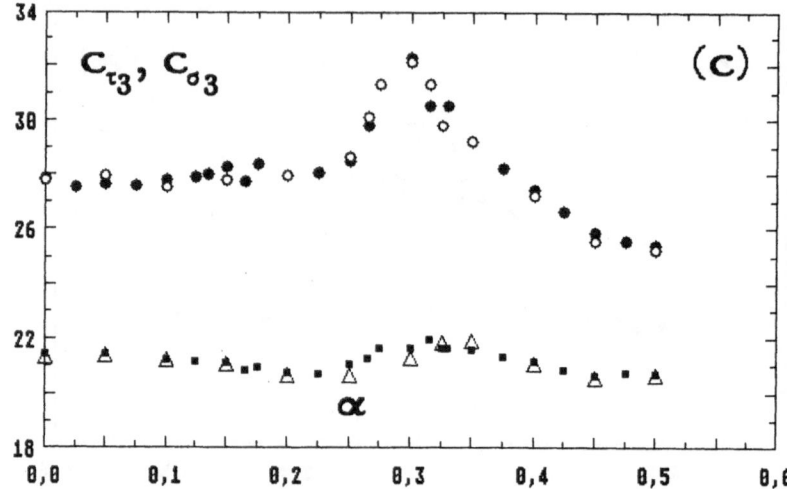

Fig. 4.13. a: Isothermal compressibility in units of $10^{-5}\mathrm{bar}^{-1}$ as a function of applied strain $\sim\alpha$: (◆) $N = 50$, (○) $N = 98$; **b:** as but for isostress expansivity in units of $10^{-4}\mathrm{K}^{-1}$; **c:** as a but for isostress heat capacity; also shown are results for the isostrain heat capacity from eq. (4.18) (■) and eq. (4.20) (△).

strain ensemble MC calculation. This procedure relies on the equivalence of different ensembles for sufficiently large systems [86].

Both directly and indirectly computed $c_{\sigma3}$ plotted in Fig. 4.13c as functions of α, agree closely. Agreement is particularly striking in the range $0.15 < \alpha < 0.40$, where $c_{\sigma3}$ exhibits a shallow minimum ($\alpha \approx 0.225$) and a small maximum ($\alpha \approx 0.325$). The comparison confirms internal consistency of the isostress-isostrain ensemble MC results for $c_{\tau3}$, $\varkappa_T$ and $\gamma_{\tau3}$ and demonstrates convincingly the equivalence between the two ensembles employed. In addition, no system size effect could be detected even in the immediate vicinity of the phase transition [201].

Transition Point Fluctuations. The origin of the previously discussed dependence of fluctuations on applied strain in the neighborhood of a phase transition may be discussed in the context of the so-called "scaling hypothesis" which can be made for thermodynamic potentials [202]. In the immediate vicinity of a phase transition it is generally possible to split the characteristic thermodynamic potential formally into a regular and into a singular part. The scaling hypothesis interprets the latter as a generalized homogeneous function, that is asymptotically in the present case (as the transition is approached)

$$G_s(\lambda^{a_1}x_1, \lambda^{a_2}x_2, \ldots, \lambda^{a_n}x_n) = \lambda\, G_s(x_1, x_2, \ldots, x_n) \qquad (4.21)$$

where G_s is the singular part of the thermodynamic potential of the isostress-isostrain ensemble, G [64], $\lambda \geq 0$ and $\{a_i\}$ represents the set of unspecified "scaling powers". Certain a_i are related by means of simple equations known as scaling laws [203].

To establish scaling laws analytically natural variables $\{x_i\}$ of G have to be chosen in a suitable way. It can be shown, that the $\{x_i\}$ must bear a certain geometric relation to the coexistence hypersurface at the critical point [202] such that the $\{x_i\}$ are related simultaneously to the usual thermodynamic variables [204], say T and P (for the special case of a homogeneous classical fluid; for magnets see [205]). From the properties of generalized homogeneous functions comprehensively discussed by Hankey and Stanley [202] it then follows in general that thermodynamic quantities may diverge at a transition point according to some power law $\sim |x|^{\nu'_i}$, if the transition is approached along any x_i-axis. The set of critical exponents $\{\nu'_i\}$ is related to $\{a_i\}$ in a way that permits formulation of scaling laws in terms of (theoretically or experimentally) accessible critical exponents. Unfortunately, in the present case the coexistence hypersurface is determined by N, T, τ_3 and σ'_3 and unknown because of its horrendous complexity. It is therefore not possible to define $\{x_i\}$. However, if the phase transition is not approached along any x_i-axis but, say along the σ_4- (i.e. α-) axis, power law divergences should still occur, but the new set of critical exponents $\{\nu_i\}$ is not obviously related to $\{a_i\}$ thereby prohibiting a test of scaling laws. Since the focus in this section is predominantly on localizing a shear melting point this latter circumstance is of little concern.

It then seems sensible to assume a power law divergence of $c_{\tau 3}$, $\varkappa_T$ and $\gamma_{\tau 3}$ as functions of α because the three quantities can be represented in terms of derivatives of G which depends explicitly on α via σ_4 [64]. The critical melting strain α_c (i.e. the shear melting point) is then obtained from a fit of $a(\alpha - \alpha_c)^\nu$, $\alpha > \alpha_c$ to the MC results. Since the plots in Fig. 4.13 suggest that the divergence occurs in the same range of α regardless of the quantity in question, the power law is fitted only to c_{τ_3} which seems to be somewhat less affected by statistical noise. A reasonable fit is obtained with $\alpha_c = 0.28$ ($\nu = -0.11$, $a = 21.1\,\mathrm{JK^{-1}mol^{-1}}$); these data are verified in [201] by comparison with the x_T-curve in Fig. 4.13a.

The results discussed so far, in particular lack of latent heat release now seem to justify description of shear melting as a second order phase transition. This is quite unusual because one deals with a solid-liquid transition which, according to Landau and Lifshitz [38] should in general not be of second but of first order. According to these authors, second order phase transitions can occur only if neighboring thermodynamic states are very similar. Such a situation arises in particular for solid-liquid phase transitions where crystal structure changes from, say fcc to bcc (body-centered cubic). It also applies to shear melting in the isostress-isostrain ensemble. The similarity becomes immediately apparent if one keeps in mind that the system passes through an infinite number of intermediate equilibrium states of infinitesimally different degrees of disorder (see Sect. 4.3.2. Therefore, if melting is driven by shear strain (instead of temperature), the phenomenological prerequisites for a second order phase transition as posed by Landau and Lifshitz are met. However, the involvement of solid- and liquidlike phases in second order phase transitions remains a striking feature and may have significant implications with regard to material properties of thin confined films.

It is also interesting to note that the order of the phase transition may depend on the thickness of the vicinal phase. While computer simulations for thin vicinal phases as described here give clear evidence of a second order shear melting transition, a first order transition will occur in the bulk ([38], p.139 in [194]). Since vicinal phases become increasingly bulklike as h grows, it seems reasonable to expect a crossover from a second to a first order phase transition with increasing h. Preliminary isostress-isostrain ensemble MC calculations indicate that a crossover point may be located somewhere between two- and five-layers vicinal phases [206].

4.3.3 Critical Evaluation of the Relation Between Computer Simulation and Laboratory Experiment

In Sect. 4.3 MC calculations are presented which aim at a more detailed understanding of shear melting. In particular MC calculations in the grand canonical ensemble support a mechanism that may account for the experimentally observed critical shear stress required to initiate sliding of solid surfaces separated by a molecularly thin film. It should be borne in mind, however, that the idealized model I employed here throughout differs in several respects from the actual experiments.

First, the structure of the walls is commensurate with that of the bulk rare-gas solid, thus promoting epitaxial growth of solidlike vicinal phases, whereas the structure of the mica sheets used in the experiments (see Sect. 4.3.1) bears no particular relation to that of bulk organic solids. X-ray investigations [207] demonstrate formation of solidlike aequous layers between silicate layers of tetrahedrally substituted 2:1 phyllosilicates. However, whether the structurally related mica surfaces used in the dynamic SFA experiments can induce "freezing" of organic liquids remains to be seen although plots in Fig. 4.6 suggest this notion (see also [187]).

Second, and perhaps more important, in the grand canonical ensemble the walls are held a fixed distance apart as they are slid. This leads to concomitant drainage from the vicinal phase. As pointed out in Sect. 4.3.1 in the corresponding experiments the normal force on the mica coated cylinders is maintained while they are slid rather than the interwall separation [189]. Under certain conditions drainage is not observed experimentally upon sliding, which implies that the number of particles remains fixed, even though the separation between the cylinders may vary.

Drainage is prohibited in the isostress-isostrain ensemble on account of the choice of state variables (N = const.). The most significant feature of shear melting in the isostress-isostrain ensemble is its apparently continuous nature, that is it occurs as a second order solid-liquid phase transition. However, neither the grand canonical nor the isostress-isostrain ensemble is natural for the dynamical SFA experiment, in which the walls are slid at constant T with the vicinal phase in contact with bulk fluid, that is with μ fixed. In addition, the walls may move simultaneously in x- and y-direction as they are slid. Thus, to simulate shear melting under constraints more closely resembling conditions under which the SFA operates, one should employ some sort of grand ensemble in which μ and T are fixed parameters. The question remains: which additional independent thermodynamic state variables should be taken to specify the macrostate of the vicinal phase, analogous to complementary strains σ'_3? The ideal choice of variables would of course be those actually controlled by the SFA. However, in the real apparatus the walls are attached to springs, which in turn are connected to points whose positions are manipulated. Thus, neither the stresses nor the strains on the vicinal phase itself are directly controlled, but rather the *strains* on the *composite* system comprising the vicinal phase plus attached springs.

Lupkowski and van Swol [208] have applied a non-equilibrium grand canonical mixed MC-MD technique to a composite-system of model I. The upper wall is subject to a constant external pressure $(-\tau_3)$ and is attached to a fixed spring which exerts a stress $\tau_4 = -K\Delta x$, where Δx is the displacement from equilibrium extension. The lower wall is translated in the x-direction at uniform speed. The shear strain in y-direction is apparently held fixed. Although this model mimics the SFA more closely than the isostress-isostrain ensemble MC method, it still falls short in that the strain σ_5 and the stress τ_3 are fixed. More important, however, is again (see comments about NEMD methods in Sect. 4.3.2) the disparity between time scales involved in the real SFA experiment and this simulation. In order to perceive appreciable relative movements of the walls during the longest simulation periods (ca. 100 ps), the wall speed must exceed that typical of the SFA by 9 or 10 orders of magnitude. Furthermore the dynamic response of the upper wall depends on its assumed mass which is arbitrarily set equal to the mass of vicinal phase atoms.

For the latter reasons it would be preferable to employ what might be called a "grand isostress" MC method, where μ, T, τ_3, τ_4, τ_5, σ_1 and σ_2 are fixed parameters instead of the dynamic composite-system approach of Lupkowski and van Swol [208]. For a given setting of laboratory-fixed points of attachment (or equilibrium positions) of the springs, the stresses acting on the vicinal phase can be determined from a knowledge of the isothermal elastic constants $\{c_{ij}\}_{i,j=1}^{6}$ of the (assumed Hookean) springs and the measured strains (extensions) of the springs. For example, in case the matrix **c** were diagonal, then the "real" experimental points μ, T, $c_{33}\Delta z$, $c_{11}\Delta x$, $c_{22}\Delta y$, σ_1, σ_2 would map onto the virtual (i.e. simulated) points μ, T, τ_3, τ_4, τ_5, σ_1 and σ_2 and vice versa. Work along those lines is in progress [209].

Thompson and Robbins also investigated shear melting by NEMD [193,196]. In their model the walls have inherent structure corresponding to the (111) plane of the fcc lattice. The vicinal phase has slit-pore geometry and the walls are held together by a constant normal pressure. Attached to the top wall is a spring connected to a stage moving at constant velocity in the x-direction. By varying the speed Thompson and Robbins observe a critical speed similar to the dynamic SFA experiments (see Fig. 4.6). Below this critical speed the vicinal phase tends to "freeze" and "melt" periodically whereas above that speed freezing is inhibited. However, the remarks concerning disparity of experimental and simulational time scales apply here, too.

5 Time-Dependent Properties

5.1 Preliminary Remarks

So far in this article only equilibrium (i. e. static) properties of vicinal phases are discussed. This regards mechanical properties such as various stress tensor elements, compressibility and expansivity, thermodynamic quantities like heat capacities but also structural aspects such as layering and lateral order. It is obvious that in particular the latter two and their dependence on thermodynamic quantities and geometric constraints are interrelated with time-dependent (i. e. dynamic) phenomena such as relaxation or transport. In order to gain a broader understanding of vicinal phases it is therefore worthwhile to devote this section to aspects of dynamic phenomena.

From a computational point of view dynamic phenomena can best be investigated within the framework of EMD or NEMD methods. Both techniques allow one to pursue particle trajectories on a physical time scale because they are based on Newton's equation. MC, on the other hand, generates a sequence of configurations which may also be interpreted temporally [59] due to the Chapman-Kolmogoroff equation which governs the time evolution of Markov processes [53] (see Sect. 2.3). However, the Chapman-Kolmogoroff equation is a master equation involving only a stochastic time. Therefore, it lacks a physically significant time scale (see Sect. 2.2.3 in [59]). Nevertheless, MC can be useful in studies of relaxation phenomena for which the physically correct short time dynamics is irrelevant. This will be shown in Sect. 5.3 where a combination of EMD and MC is employed to study anomalous self-diffusion in the vicinity of solidification of vicinal phases.

In principle, a conceptual problem arises when EMD is applied to vicinal phases. As has been pointed out before (see Sect. 3.4, 4.3.2) MC ensembles are selected with special emphasis on complementary experiments. The discussion of EMD in Sect. 2.2, on the other hand, shows that EMD in its simplest version based upon eq. (2.1) is representative of the microcanonical ensemble which is not directly related to experiments on vicinal phases. However, as before with MC it is most interesting to mimic experimental situations in EMD to the largest possible extent. This can be achieved by employing the postulate of equivalence of statistical-physical ensembles. For example, instead of fixing μ, V and T in the grand canonical ensemble one may alternatively chose $N = \langle N(\mu) \rangle$, V and $\langle E \rangle$ in EMD as thermodynamic state parameters, where $\langle .... \rangle$ denotes grand canonical ensemble averages.

The postulate of equivalence of statistical-physical ensembles relies largely on ergodicity of phase (configuration) space trajectories [8,41]. Ergodicity, on the other hand, is not immediately guaranteed. In practice, one may encounter cases where MC and EMD converge to different "equilibrium" states although equivalent state parameters are chosen.

Such a situation arises, for instance, if the vicinal phase tends to solidify. Starting from a random configuration with an arbitrary number of particles in grand canonical ensemble MC will eventually allow for solidification of the vicinal phase by virtue of addition/removal steps in the modified Metropolis algorithm (see Sect. 2.3.1). Taking $N = \langle N(\mu) \rangle$ as input for an EMD run and placing particles at random in the imaginary simulation cell, on the other hand, will most likely cause several particles to get "stuck" at interstitial sites even after equilibration. These sites are surrounded by large energy barriers due to the otherwise regular structure of the vicinal lattice (see, for instance, Fig. 6(d) in [82]). Interstitially captivated particles cannot easily surmount these barriers. "Equilibrium" properties obtained from EMD will then be characterized by a certain disparity compared with corresponding results from grand canonical ensemble MC.

To justify the postulate of ensemble equivalence from a practical perspective it is generally advisable to compare static properties accessible to both EMD and MC. Ergodicity problems in EMD of the sort just described can best be circumvented if grand canonical ensemble MC is utilized during the EMD equilibration period.

The advantage of MD over MC results from direct accessibility of time-dependent properties. Quantitatively, time-dependence of various properties can best be discussed in terms of autocorrelation functions

$$C(t) = \frac{\langle \varphi(0)\,\varphi(t)\rangle}{\langle \varphi(0)^2\rangle} \qquad (5.1)$$

where φ denotes any dynamical variable (i.e. velocity, force, stress tensor etc.). According to eq. (5.1) one has $C(0) = 1$ and $C(\infty) = 0$ if $\langle \varphi \rangle = 0$. While $C(t)$ permits detailed insight into microscopic dynamics, it is often desirable for reasons of comparison with experiments to "summarize" the detailed information buried in $C(t)$ in terms of a transport coefficient. The fluctuation-dissipation theorem [210-212] allows one in general to establish such a relation via

$$Y \sim \int\limits_0^\infty C(t)\,dt \qquad (5.2)$$

which is known as Green-Kubo integral [210-212]. Y is a macroscopically introduced transport coefficient. To establish a quantitative link between Y and $C(t)$ one needs to know the relevant macroscopic transport equation.

While this is formally a rather straightforward procedure in the bulk, complications arise for vicinal phases on account of inhomogeneity and anisotropy. To highlight these complications, the present chapter is devoted to a discussion of self-diffusion in vicinal phases.

5.2 Self-Diffusion in Vicinal Phases

5.2.1 Macroscopic and Microscopic Description of Self-Diffusion

Perhaps the simplest transport process one may envision in vicinal phases is mass transport. In pure systems (which are of exclusive concern here) self-diffusion is the only possible realization of mass transport at thermodynamic equilibrium. Self-diffusion in general obeys Fick's (first and second) law(s), namely

$$\frac{\partial c(\mathbf{r},t)}{\partial t} = \nabla \cdot \mathbf{D} \cdot \nabla c(\mathbf{r},t) \tag{5.3}$$

where $c(\mathbf{r},t)$ is the concentration of particles at $\mathbf{r}$ and t; $\mathbf{D}$ denotes the diffusion tensor which in general is position- and time-dependent [213,214]. In the hydrodynamic limit [215], however, one may take $\mathbf{D}$ to be independent of $\mathbf{r}$ and t. According to the slit-pore geometry $\mathbf{D}$ is a second rank tensor and diagonal if the principal directions are along the crystal axes. If in-plane isotropy (i.e. (x,y)-plane) is preserved $\mathbf{D}$ is composed of two non-vanishing elements, namely

$$D_{\parallel} = \frac{1}{2}\left(D_{xx} + D_{yy}\right) \tag{5.4a}$$

$$D_{\perp} = D_{zz} \tag{5.4b}$$

From the above assumptions it follows that the diffusion equation (5.3) is separable in laboratory Cartesian coordinates. Since periodic boundary conditions are imposed on the (x,y)-plane (see Sect. 3.3) the solution of eq. (5.3) in x- and y-directions proceeds as in homogeneous, isotropic bulk phases [216]. It is then possible to link macroscopically defined diffusion tensor elements to microscopic expressions. It is easy to show that by analogy with eq. (5.1) and (5.2)

$$D_{\alpha\alpha} = \int_0^\infty \langle \dot{\alpha}(0)\,\dot{\alpha}(t)\rangle\, dt \tag{5.5a}$$

provided the integrand decays fast enough where α is the α-component of $\mathbf{r}(\alpha = x, y)$. It can also be shown that (see p. 201 in [52], p. 515 in [8])

$$2D_{\alpha\alpha}\, t = \lim_{t \to \infty} \langle \Delta\alpha^2(t)\rangle \tag{5.5b}$$

where

$$\langle \Delta \alpha^2(t) \rangle = \langle [\alpha(0) - \alpha(t)]^2 \rangle \qquad (5.5c)$$

is the mean square displacement (MSD) in α-direction. The integrand in eq. (5.5a) is known as velocity autocorrelation function to which the acronym VACF is assigned. Both MSD and VACF are readily accessible in EMD. Since particles can be assumed to be statistically independent one may use

$$\langle \dot{\alpha}(0) \dot{\alpha}(t_k) \rangle = (NL)^{-1} \sum_{l=1}^{L} \sum_{i=1}^{N} \dot{\alpha}_i(t_l) \dot{\alpha}_i(t_l + t_k) \qquad (5.6a)$$

$$\langle [\alpha(0) - \alpha(t_k)]^2 \rangle = (NL)^{-1} \sum_{l=1}^{L} \sum_{i=1}^{N} [\alpha_i(t_l) - \alpha_i(t_l + t_k)]^2 \qquad (5.6b)$$

and take α_i, $\dot{\alpha}_i$ from the sequence of EMD configurations stored at fixed time intervals $\Delta t = t_{l+1} - t_l$. The averaging is performed over N statistically independent particles and L "time origins". Both VACF and MSD can be expected to have typical error bars of $\leq 3\%$ if the product $NL \geq 10^5$.

Solution of the Diffusion Equation in Z-Direction. While eqs. (5.6a) and (5.6b) are valid for $\alpha = x,y,z$, eqs. (5.5a) and (5.5b) are not in general applicable to diffusion in z-direction. Relations similar to eqs. (5.5a) and (5.5b) can be established by solving the diffusion equation in z-direction. This can be done provided the boundary conditions are properly altered due to confinement by the walls. Following Hall and Ross [217] boundary and initial conditions can be stated here as

$$c(z,0) = \delta(z - z_0) \qquad (5.7a)$$

$$\left. \frac{\partial c(z,t)}{\partial z} \right|_{z=0,h} = 0 \quad , \quad t > 0 \qquad (5.7b)$$

Eq. (5.7b) is known as no-flux boundary condition and is fulfilled rigorously for all the models studied here on account of the hard-core background potential in eqs. (3.5a), (3.5b), (3.7) and (3.8).

Hall and Ross [217] obtain a series solution of the diffusion equation subject to the above boundary conditions which can be written as

$$c(z',z_0,t) = \frac{1}{h} + \frac{2}{h} \sum_{n=1}^{\infty} \exp(-n^2\pi^2 D_\perp t/h^2) \cos\left(\frac{n\pi(z'+z_0)}{h}\right)\cos\left(\frac{n\pi z_0}{h}\right)$$

(5.8)

where $z' = z - z_0$. To yield the conditional probability $G_{s,z}$ of finding a diffusing atom between z' and $z' + dz'$ at time t if it was at the origin at $t = 0$ one needs to average eq. (5.8) over all possible origins z_0. One obtains

$$G_{s,z}(z',t) = \frac{h \pm z'}{h^2} + \sum_{n=1}^{\infty} \exp(-n^2\pi^2 D_\perp t/h^2)\left[\frac{h \pm z'}{h^2}\cos(n\pi z'/h)\right.$$

(5.9)

$$\left. \pm \frac{1}{n\pi h}\sin(n\pi z'/h)\right]$$

where "+" refers to $z' < 0$ and "-" to $z' > 0$. It is worth noting that $G_{s,z}(z',t)$ does not satisfy the diffusion equation on account of the averaging process; $c(z',z_0,t)$, on the other hand, is a solution of the diffusion equation which can easily be verified by inserting eq. (5.8) into eq. (5.3). With the aid of eq. (5.9) it is straightforward to compute the MSD in z-direction as

$$\langle \Delta z^2(t)\rangle = \int_{-h}^{h} z'^2 G_{s,z}(z',t)\, dz' = \frac{h^2}{6} - \frac{16h^2}{\pi^4}\sum_{n=1}^{\infty}(2n-1)^{-4}$$

(5.10)

$$\times \exp\left[-(2n-1)\pi D_\perp t/h\right]$$

From a fit of eq. (5.10) to the EMD generated MSD (eq. (5.6b)) $D_\perp$ can be determined. Fig. 5.1 shows a plot of the EMD generated MSD and a fit to it using the first ten terms of the sum in eq. (5.10).

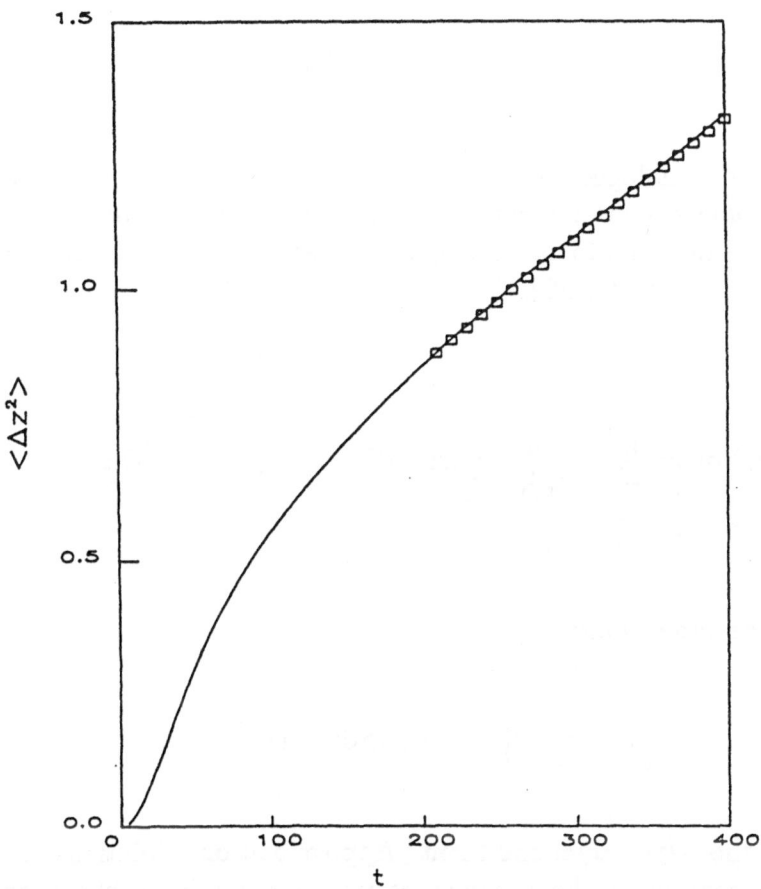

Fig. 5.1. Plot of mean square displacement normal to the walls for model I at $-\mu^* = 10.0$, $T^* = 1.00$, $h^* = 4.80$ and $\alpha = 0.5$. The open squares are from the least-squares fit of eq. (5.10), using $n = 10$ terms. Units are $10^{-20}\,\mathrm{m}^2$ ($\langle\Delta z^2\rangle$) and $10^{-14}\mathrm{s}$ (t).

It is instructive to examine the long time and "thick" vicinal phase limits of eq. (5.10). According to the Weierstrass M test (p. 363 in [218]), the infinite series in eq. (5.10) converges uniformly for $t \geq 0$. Therefore, the limit $t \to \infty$ may be interchanged with the summation sign. It follows that the MSD tends to a constant $h^2/6$ at long times

[219]. Again by means of the Weierstrass M test, one can show that the series in eq. (5.10) converges uniformly as h ⟶ ∞; the Maclaurin series expansion of exponentials likewise converges uniformly. Thus one finds from eq. (5.10) [219]

$$\lim_{h \to \infty} \langle \Delta z^2(t) \rangle = 2 D_\perp t \tag{5.11}$$

which is the analogue of eq. (5.5b) and represents the result for an infinite homogeneous medium. In the limit h ⟶ ∞ one may also assume the medium to be isotropic so that $D_\parallel = D_\perp$. It may furthermore be shown that [219]

$$\int_0^\tau \langle \dot{z}(0) \dot{z}(t) \rangle \, dt = \frac{h^2}{12\tau} - \frac{8h^2}{\pi^4 \tau} \sum_{n=1}^\infty (2n-1)^{-4} \exp\left[-(2n-1)^2 \pi^2 D_\perp \tau / h^2\right] \tag{5.12}$$

and by the same token

$$\lim_{h \to \infty} \lim_{\tau \to \infty} \int_0^\tau \langle \dot{z}(0) \dot{z}(t) \rangle \, dt \simeq D_\perp \tag{5.13}$$

Test of the Hydrodynamic-Limit Approximation. Solutions of the diffusion equation in the previous section are explicitly based on the hydrodynamic-limit approximation, i.e. diffusion tensor elements are constants [215]. While this assumption is very likely to be uncritical in eqs. (5.5a) or (5.5b) for the majority of thermodynamic states considered due to the semiinfinite extension of vicinal phases in the (x,y)-plane, some caution is advised for diffusion perpendicular to the walls if h gets very small so that only a molecularly thin, strongly layered film is accomodated.

However, it will be shown in Sect. 5.3 that self-diffusion in the (x,y)-plane can be non-Brownian for certain thermodynamic states thereby invalidating eqs. (5.5a) and (5.5b).

One may test the validity of the hydrodynamic-limit approximation by considering an explicitly time- and position-dependent diffusion tensor. It is introduced by considering the self-part of the intermediate scattering function [52]

$$F_s(\mathbf{k}, t) = \left\langle \exp\left[-i\mathbf{k} \cdot \mathbf{r}(0)\right] \exp\left[i\mathbf{k} \cdot \mathbf{r}(t)\right] \right\rangle \tag{5.14}$$

which represents the autocorrelation function of Fourier components of the local density. $F_s(\mathbf{k}, t)$ can in principle be measured in incoherent neutron scattering experiments [65]. It also satisfies Volterra's equation

$$\frac{dF_s(\mathbf{k}, t)}{dt} = -\int_0^t K(\mathbf{k}, t - t') \, F_s(\mathbf{k}, t') \, dt' \tag{5.15}$$

The kernel K (referred to as "memory function") is then given by [219]

$$K(\mathbf{k}, t) = \mathbf{k} \cdot \boldsymbol{D}(\mathbf{k}, t) \cdot \mathbf{k} \tag{5.16}$$

where $\boldsymbol{D}(\mathbf{k}, t)$ is the position- and time-dependent diffusion tensor. Employing the projection operator formalism [8,220] one can show (see Appendix in [219]) that

$$\boldsymbol{D}(\mathbf{k}, t) \longrightarrow \langle \dot{\mathbf{r}}(0) \, \dot{\mathbf{r}}(t) \rangle + O(k^2) \tag{5.17}$$

Since it is most interesting to test the validity of the hydrodynamic approach in z-direction, $\mathbf{k} = k\hat{\mathbf{e}}_z$ is chosen. Then, from eqs. (5.16) and (5.17) it follows that

$$\lim_{k \to \infty} k^{-2} K(k\hat{\mathbf{e}}_z, t) = \lim_{k \to \infty} D_{zz}(k\hat{\mathbf{e}}_z, t) = \langle \dot{z}(0) \, \dot{z}(t) \rangle = C_\perp(t) \tag{5.18}$$

Both $C_\perp(t)$ and $D_{zz}(k\hat{e}_z,t)$ can be computed in the same EMD calculation. Therefore, eq. (5.15) needs to be solved for the numerical $F_s(k\hat{e}_z,t)$ which is done by the algorithm described by Berne and

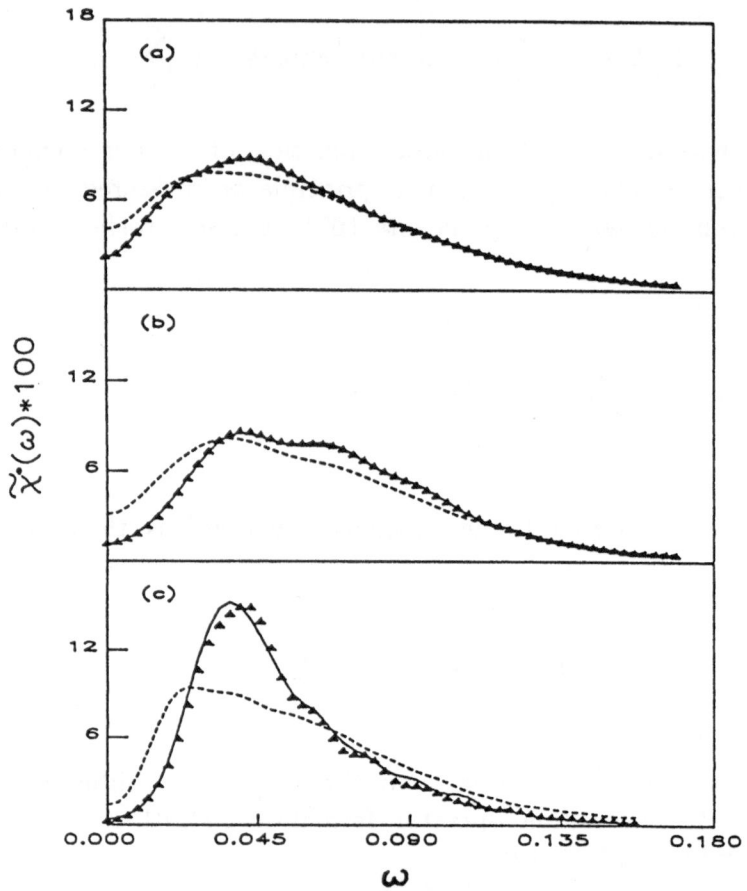

Fig. 5.2. Fourier transforms of $D_{zz}(k\hat{e}_z,t)$ (—), $C_\perp(t)$ (▲) and $C_\parallel(t)$ (----) for **a**: $h^* = 4.8$, **b**: $h^* = 3.7$, **c**: $h^* = 2.9$.

Harp [221]. Fig. 5.2 shows plots of Fourier transforms of $C_\perp(t)$ and $D_{zz}(k\hat{e}_z,t)$ for three different values of $k\hat{e}_z = 2\pi h^{-1}$. As can be seen from that figure the two quantities agree for all but the smallest $h^* = 2.90$. It is therefore concluded that the hydrodynamic approach breaks down for very thin vicinal films of less than three layers of fluid.

5.2.2   Inhomogeneity and Anisotropy

Perhaps the most significant structural feature of vicinal phases concerns inhomogeneity which becomes manifest as layering (see Sect. 3.4.2). If one analyzes the period of time τ_D an atom spends in a particular layer it turns out that for a sufficiently large fraction of vicinal atoms (see Table VIII in [229]) τ_D exceeds the correlation time

$$\tau_C = \lim_{\tau \to \infty} \int_0^\tau C_\parallel(t)\, dt \qquad (5.19)$$

where $C_\parallel(t)$ denotes the VACF of lateral diffusion. According to the relation $\tau_D > \tau_C$ it is sensible to define a diffusion coefficient

$$D_\parallel^{(i)} = \frac{1}{2}\left[D_{xx}(z^{(i)}) + D_{yy}(z^{(i)})\right] \qquad (5.20)$$

where $D_{\alpha\alpha}(z^{(i)})$, $\alpha = x, y$ denotes a diffusion coefficient defined analogously to eqs. (5.5a) or (5.5b) with the additional constraint that the sums over i in eqs. (5.6a) or (5.6b) extend only over a certain sub-group of atoms. This subgroup is associated with atoms originally starting out in layer i centered on $z^{(i)}$ and remaining in that layer for times $\tau = \tau_C$. In Fig. 5.3 results are displayed for model I at $-\mu^* = 10.0$, $T^* = 1.0$. Regardless of h diffusion is smallest in the contact layer due to friction experienced by atoms moving next to the (rigidly fixed) wall atoms. Diffusion increases toward the center of the vicinal phase due to decreasing friction. For each layer $D_\parallel^{(i)}$ is an oscillatory function of h. It is reasoned in [219] that this effect is due to variations in pore average density but also to changes in the layered structure perpendicular to the walls (see Fig. 6 in [219]).

If vicinal phases are sufficiently thin inhomogeneity is accompanied by anisotropy. Anisotropy becomes manifest here as different mobility of diffusing atoms in directions parallel and perpendicular to the walls. A measure of anisotropy is the ratio $D_\perp/D_\parallel$ plotted in Fig. 5.4 for model I. Clearly, $D_\perp/D_\parallel = 1$ if diffusion is isotropic. From the plot in Fig. 5.4 it is evident that this limiting value will be assumed if $h \to \infty$ which makes physical sense because it is shown in Fig. 3.5 that the homogeneous bulklike region of the vicinal phase tends to

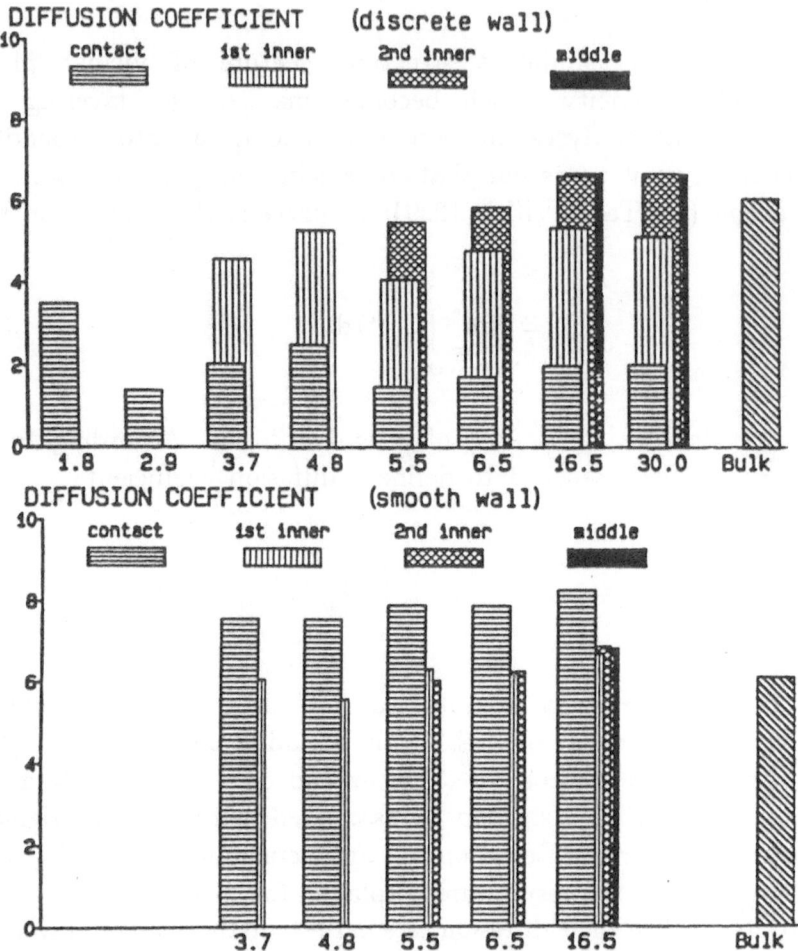

Fig. 5.3. Dependence of local transverse component of the diffusion tensor on distance of vicinal layer from the walls of model I (upper plot), as a function of separation between the walls; the lower plot shows the corresponding data for model II.

dominate over the inhomogeneous, layered region as h increases. However, the value $D_\perp^*/D_\parallel^* = 0.93$ also signifies that even at $h^* = 30.0$ ($\approx 100\,\text{Å}$) the vicinal phase lacks complete isotropy (i.e. not truly has bulk phase characteristics). This notion is supported by a slightly higher value of $D_\parallel^{(i)}$ for the homogeneous middle section compared with the corresponding bulk value (see Fig. 5.3). As h decreases $D_\perp/D_\parallel$

goes to zero on account of the increasingly restricted diffusivity perpendicular to the walls.

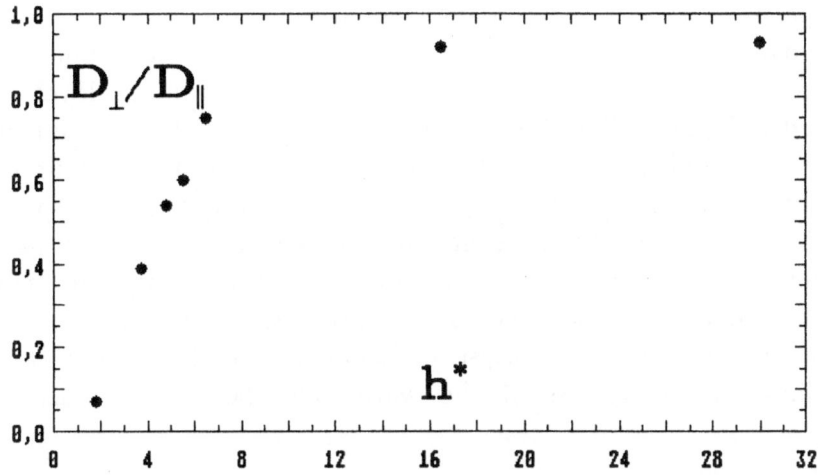

Fig. 5.4. The ratio of normal and transverse components of the diffusion tensor for fluidlike vicinal phases as a function of distance between the walls at $-\mu^* = 10.0$, $T^* = 1.00$.

5.2.3   The Influence of Wall Structure

Besides inhomogeneity and anisotropy which are more generic features of vicinal phases, a more detailed influence of the walls may be expected. In Sect. 3.4.2, for instance, it is demonstrated that only model I allows for solidification of vicinal phases. Models II and III prohibit solidification because of their complete lack of structure. However, even details of structured walls are important. This follows from results for model IV which possesses some sort of infinitesimal random structure but is shown in [51] to be even counter-productive in the formation of solidlike vicinal phases. It is therefore sensible to expect similarly strong influences of wall structure on self-diffusion (and, perhaps, other dynamic properties, too).

This is confirmed by the plots of $D_{\parallel}^{(i)}$ for model II in Fig. 5.3. The most striking feature about this plot is the reversed dependence of $D_{\parallel}^{(i)}$ with increasing distance from the wall in comparison with results for model I also presented in that figure. For model II atoms

110

are always diffusing fastest in the contact layer. Physically this behavior reflects lack of lateral friction which can easily be understood because Φ_{pw} in eq. (3.7) depends only on an atom's relative position with respect to both walls. There is only little dependence of $D_{\parallel}^{(i)}$ on i for all but the contact layer irrespective of h.

This overall behavior is very similar to model III for which the MSD's for lateral diffusion in the contact and first inner layer are presented in Fig. 5.5. The MSD's pertain to a vicinal phase comprising four layers in which the contact layers completely wet the walls due to the vanishing range of Φ_{pw} in eq. (3.8). However, due to the complete lack of roughness atoms in the contact layer are again significantly faster than those travelling in inner layers. The same effect is observed if model IV is employed although the deviation between contact and inner layer diffusion is diminished because of the infinitesimal roughness of the walls. This behavior is also re-

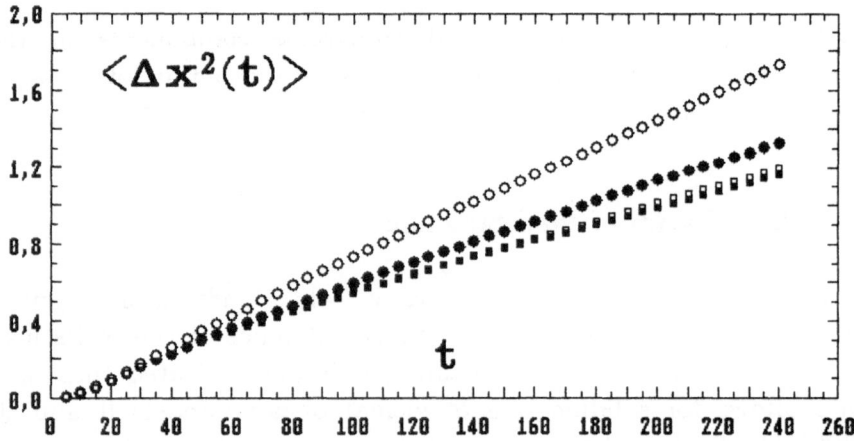

Fig. 5.5. Mean square displacement versus time within contact layer [model III ($\diamond$); model IV ($\blacklozenge$)] and inner layer [model III ($\square$); model IV ($\blacksquare$)] (see Fig. 3.6). Units see Fig. 5.1.

flected by the ratio of diffusion coefficients for model III, $D_{\parallel}$ (III), and model IV, $D_{\parallel}^{(i)}$(IV). From results displayed in Table III of [51] one finds $D_{\parallel}^{(1)}$(III)$/D_{\parallel}^{(1)}$(IV)$=1.38$ respectively $D_{\parallel}^{(2)}$(III)$/D_{\parallel}^{(2)}$(IV)$=1.00$ which also indicate that only diffusion in the contact layer is affected by details of the particle-wall interaction for these two models.

5.3   Anomalous Diffusion

5.3.1   The Role of Spatial Hindrance

In Sect. 5.2 self-diffusion is introduced as a dynamic process governed by Fick's law(s) (eq. (5.3)). Solving eq. (5.3) for atomic motion in the semiinfinite (x,y)-plane leads to a relationship between MSD and time (eq. (5.5b)). Usually this relation turns out to be linear in EMD calculations for times $t > \tau_C$ (see eq. (5.19)) which is characteristic of Brownian motion [222]. However, in cases where diffusion in the (x,y)-plane is spatially hindered, the approach to Brownian motion may be delayed. Eq. (5.5b) can then be replaced by the more general equation [213,223,224]

$$2d^{-1}D_{\alpha\alpha}\,t^d = \langle \Delta\alpha^2(t) \rangle \quad , \quad t > \tau_C \tag{5.26}$$

where $0.0 < d < 1.0$ is the fractional dimension. Diffusion processes governed by a time dependence $\sim t^d$ are therefore often referred to as "fractional Brownian motion" [223]. Note, that in the limit $d = 1.0$ eq. (5.26) reduces to ordinary Brownian motion (eq. (5.5b)). The fractional dimension may be deduced in principle from a least-squares fit to a double-logarithmic representation of eq. (5.26).

Spatial hindrance arises, for instance, in the monolayer vicinal phase studied in Sect. 4.3.2. It is pedagogically expedient to view diffusion in the monolayer film as resulting from movements of atoms through a rigid microporous medium (i.e. a regular array of cells connected by tunnels generated by the external potential field of the fixed wall atoms in model I). The diagrams of Fig. 5.6 depict the sorts of structures a diffusing atom may encounter. Fig. 5.6c illustrates the case in which the walls are precisely in registry ($\alpha = 0.0$) and the monolayer is solidlike. Every mobile atom is trapped in a cell formed by its eight near-neighbor wall atoms. On any timescale it cannot surmount the effective potential barrier blocking its migration to a neighboring cell. The tunnels are too constricted. At the other extreme (Fig. 5.6a) where the walls are completely out of registry ($\alpha = 0.5$), the barriers to migration are minimum and diffusion is facile.

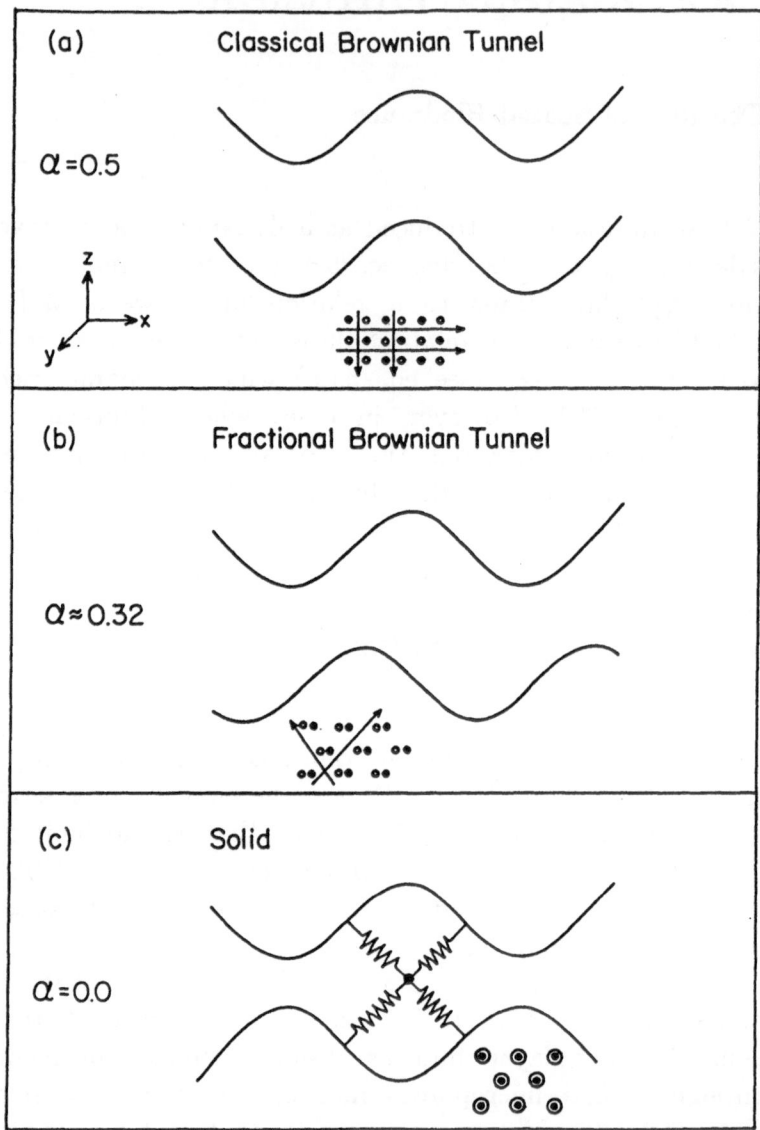

Fig. 5.6. Types of microstructures encountered by a diffusing atom. Atoms represented by open (upper wall) and filled (lower wall) circles. Arrows indicate paths of diffusing atoms and oscillatory curves projections of equipotential surfaces onto planes containing those paths.

The intermediate case (Fig. 5.6b) is most intriguing. For values of α beyond the inception of shear melting ($\alpha \gtrsim 0.3$) (see Sect. 4.3.2), the tunnels are less constricted than for $\alpha = 0.0$. At short times a mobile atom merely oscillates about its equilibrium position within its cell. By intermediate times it has begun to explore the periphery of its original cell. Through cooperative motions an atom occasionally "hops" to a neighboring cell, where it oscillates for a while before hopping again. This hopping, which is directly observed in motion pictures based on EMD trajectories begins to occur on increasingly longer timescales the closer the particular state is to the shear melting point $\alpha_c \simeq 0.28$.

5.3.2 Fractional Dimension of Self-Diffusion and Its Relation to Shear Melting

Recurrence time. The qualitative observations described in the previous section need to be cast into quantitative terms. This can be done by computing the fractional dimension $d(\alpha)$ from the double-logarithmic representation of eq. (5.26) as outlined in Sect. 5.3. It is important to note again, that eq. (5.26) is only valid for $t > \tau_C$, the correlation time defined in eq. (5.19). τ_C is a function of registry α and can be estimated from the plots in Fig. 5.7 which exhibit VACF's at various α and as functions of time. It is evident from the figure that

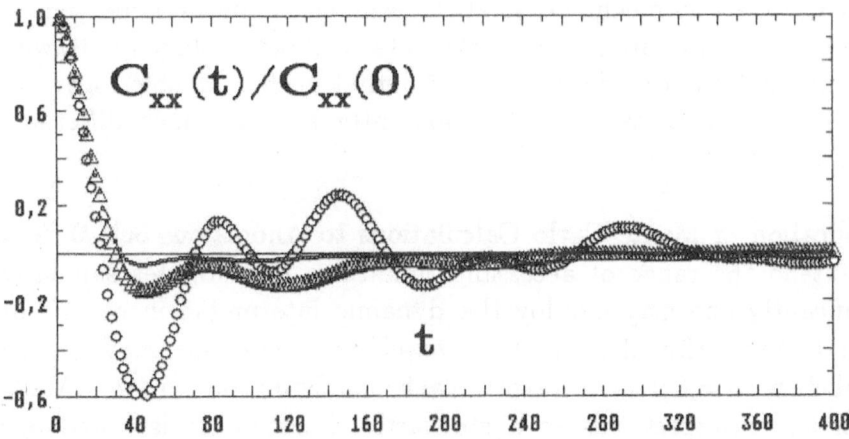

Fig. 5.7. Normalized velocity autocorrelation function in x-direction versus t (in 10^{-14}s) for a monolayer vicinal phase at $-\tau_3^* = 0.234$, $T^* = 1.00$, $N = 512$, $s^* = 25.576$. ($\square$) $\alpha = 0.50$, $h^* = 2.015$; ($\triangle$) $\alpha = 0.30$, $h^* = 1.923$; ($\lozenge$) $\alpha = 0.00$, $h^* = 1.682$.

$C_{xx}(t)$ decays to zero approximately between $2\,ps\,(\alpha=0.5)<\tau_C<2.5\,ps$ $(\alpha=0.30)$.

The structure of $C_{xx}(t)$ also reflects hopping and oscillatory motions of vicinal atoms. At $\alpha=0.5$ where the motion picture reveals virtually no oscillatory motion $C_{xx}(t)$ decays rapidly to zero. As α decreases the extent of oscillatory motion increases due to the increasingly constricted tunnels. Hence $C_{xx}(t)$ exhibits an increasingly oscillatory structure. At $\alpha=0.0$ where no hopping can occur, the structure of $C_{xx}(t)$ is highly oscillatory and decorrelation takes much longer, i.e. τ_C increases due to the very limited region of space vicinal particles can visit.

While τ_C gives a lower limit for the validity of eq. (5.26) the upper limit should be taken as large as possible for a double-logarithmic representation to be sensible. In other words, t should vary over at least two to three orders of magnitude to be sure about the linear dependence of $\ln\langle\Delta\alpha^2(t)\rangle$ on $\ln t$. In EMD, however, this is usually impossible. Since EMD is based on Newton's equation it allows for propagating modes due to thermal fluctuations. Because of periodic boundary conditions these modes may recur after a time τ_R which is determined by the system size s and the velocity of sound. Recurrence of propagating modes causes spurious correlations which render unreliable EMD generated correlation functions for $t>\tau_R$. For $s^*=25.576$ (N=512) employed here $\tau_R\approx8\,ps$ is estimated. Clearly, the ratio $\tau_R/\tau_C\approx4$ is much too small to establish a linear dependence of $\ln\langle\Delta\alpha^2(t)\rangle$ on $\ln t$ from the EMD data without ambiguity. However, it is noted from the plots in Fig. 5.8 that such a double-logarithmic plot of the EMD data yields apparently straight lines differing in slope at different α.

Application of Monte Carlo Calculations to Anomalous Self-Diffusion.
To extend the range of accessible timescales in computer simulations significantly one may employ the dynamic interpretation of MC outlined in Sect. 2.2.3 of [59]. As is explained there one may associate a label t_ν, $\nu=0,1,2,\ldots\ldots$ with each configuration in the Markov chain and interpret this as a stochastic timescale. It is noted again that the stochastic timescale does not necessarily have anything in common with the physical timescale underlying an equation of motion. This is immediately evident from the Chapman-Kolmogoroff equation [53] which does not allow for propagating modes.

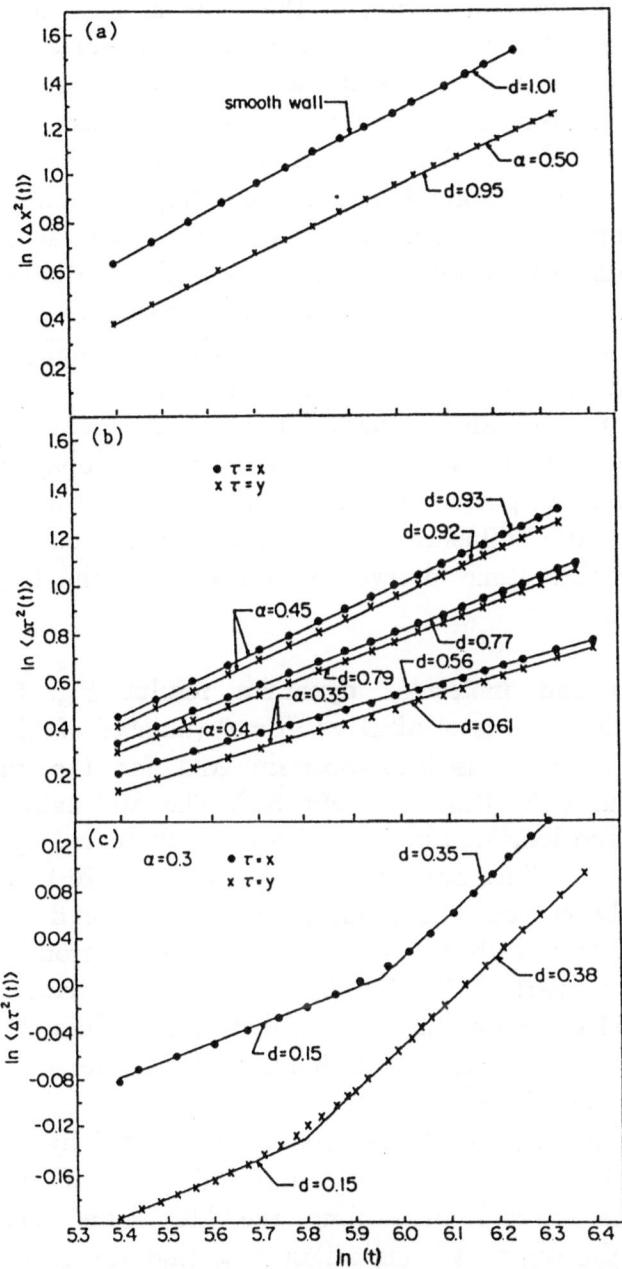

Fig. 5.8. Log of mean square displacement (units of $10^{-20}\mathrm{m}^2$) versus log of time (units of 10^{-14}s) for a monolayer vicinal phase (model I) at $T^* = 1.00$ and $-\tau_3^* = 0.234$ from EMD. $N = N_s = 512$, $s^* = 25.576$. **a**: model II (labelled "smooth wall"): $h^* = 1.923$; model I: $\alpha = 0.50$, $h^* = 2.015$. Because $\langle\Delta x^2\rangle = \langle\Delta y^2\rangle$, only $\langle\Delta x^2\rangle$ is shown. **b**: model I: $\alpha = 0.40$, $h^* = 1.996$; $\alpha = 0.35$, $h^* = 1.967$. **c**: $\alpha = 0.30$, $h^* = 1.923$ (from [224])

From the previous discussion of the plots in Fig. 5.7, however, it may be inferred that physical and stochastic timescales may become comparable for times $t > \tau_C$ as far as self-diffusion is concerned. In other words, diffusion may be described as a stochastic process after a certain period of time during which vicinal atoms have completely lost memory of their original motion. However, it is noteworthy that loss of memory is not sufficient to have $d = 1$. On account of the constricted tunnels between neighboring cells spatially non-local effects remain which may cause the t^d-dependence of the MSD. Thus, for times $t > \tau_C$ EMD and MC generated MSD's should be equivalent with respect to their time dependence.

The advantage of MC is immediately evident because recurrence effects are prohibited due to the supression of propagating modes. One may, therefore, pursue MC MSD's in principle to arbitrarily long times without any influence from spurious correlations. In practice, the length of these time intervals is restricted by the length of the Markov chain.

Monte Carlo and molecular dynamics results. Fig. 5.9 displays double-logarithmic plots of MSD vs. "time" (number of MC steps per atom) obtained from isostress-isostrain MC for the same states previously studied by EMD (see Fig. 5.8). The MC data cover both intermediate and long timescales, which are demarcated by a break in the slope. Straight lines are least-squares fits (eq. (5.26)) to the linear portions of the curves. The resulting best values of d are given in Table 5.1 together with the corresponding values from fits to the accessible linear part of EMD MSD in the intermediate time range. From Table 5.1 one notes that intermediate-time values of d between the two methods agree, which tends to validate the stochastic description of self-diffusion.

Note that MC and MD timescales can be related by requiring the intermediate-time portions of MC and MD plots of the MSD to agree. This is effected by scaling the MC "time" according to $t' = \text{const.} \times t$, the constant being adjusted so that the two coincide. It is furthermore interesting to note that the MC MSD is characterized by a larger d at longer times. The estimated times at which the break from smaller to larger d occur are listed in Table 5.1. These times correspond roughly to root mean square displacements of $0.5 \sigma \sim 1$ (lattice constant) indicating that atoms are now beginning to explore the periphery of their original cells. Hence the break in the

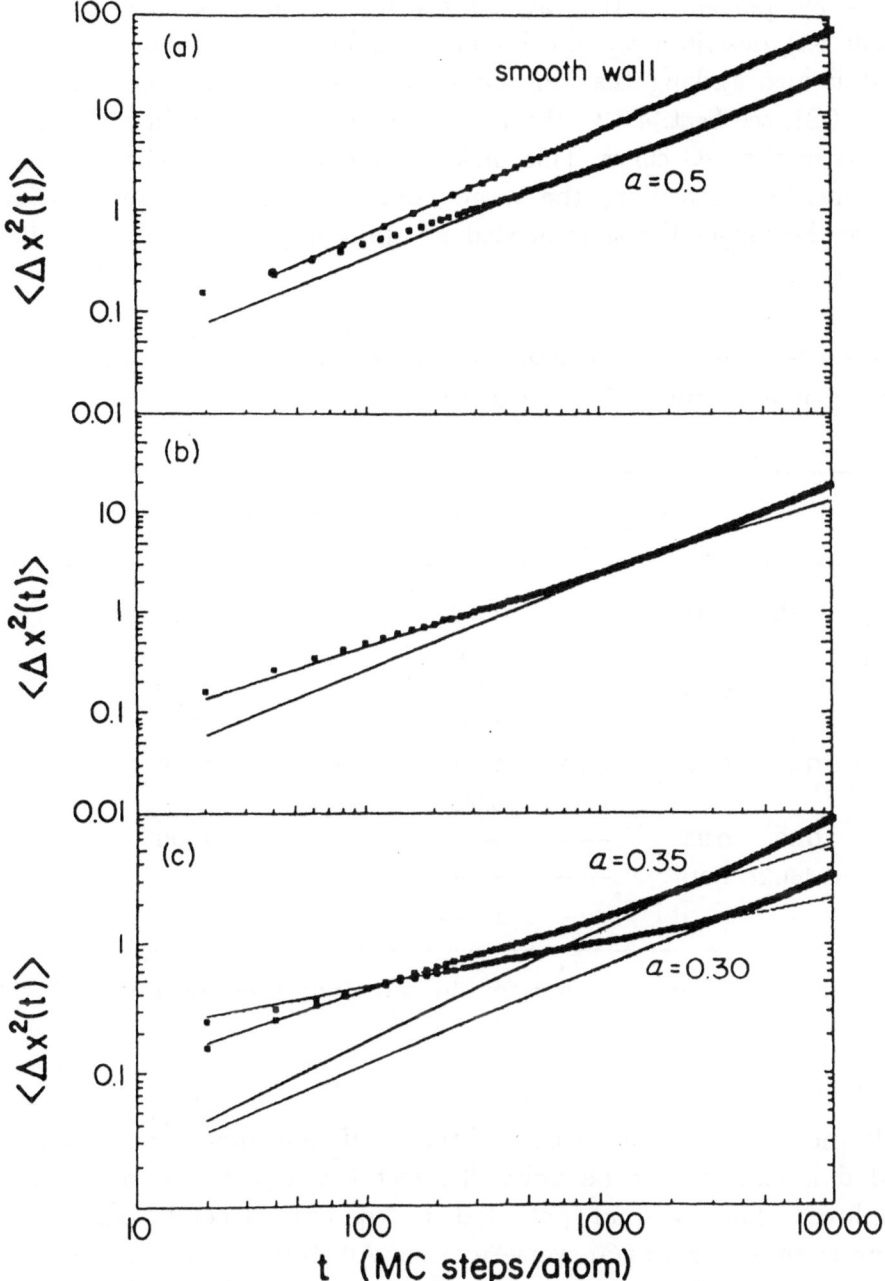

Fig. 5.9. Mean square displacement (units of $10^{-20}\mathrm{m}^2$) versus "time" (MC steps per atom) for a monolayer vicinal phase at $T^* = 1.00$ and $-\tau_3^* = 0.234$; $N = N_s = 50$, $s^* = 7.9925$. **a:** model II (labelled "smooth wall"), $\langle h^* \rangle = 1.923$; $\alpha = 0.50$, $\langle h^* \rangle = 2.015$. **b:** $\alpha = 0.40$, $\langle h^* \rangle = 1.996$. **c:** $\alpha = 0.35$, $\langle h^* \rangle = 1.967$; $\alpha = 0.30$, $\langle h^* \rangle = 1.923$ (from [224]).

long-time portion of the MC curves, for which d is close to 1 (see Table 5.1), describes random hopping among near-neighbor cells.

It is also striking that the closer α is to the shear melting point ($\alpha_c \simeq 0.28$, see Sect. 4.3.2), the greater is the time at which the break occurs in the MC curve. This makes sense in that the nearer α is to the melting transition, the more constricted are the tunnels and hence the longer the time needed for diffusing atoms to effect a hop.

Table 5.1. Temporal characteristics of diffusion in monolayer vicinal phases as a function of the registry.

α	d_{MD}		d_{MC} [a)]	t_{break}(MC steps/atom)
	x	y		
0.30	0.35	0.38	0.3 4	3400
			(0.67)	
0.35	0.56	0.61	0.5 6	2000
			(0.85)	
0.40	0.77	0.79	0.7 4	1300
			(0.94)	
0.50	0.95	——	——	1000
model II	1.01	——	——	——
			(1.05)	

[a] Numbers in parentheses refer to the long-time portion of MSD Curves in Fig. 5.9.

If the walls had no structure, then diffusion should be Brownian and d is expected to be unity. To test this hypothesis, model II is employed. The external potential in eq. (3.7) depends only on z; there is no barrier to transverse motion. An illustration of model II in the vein of Fig. 5.6 would consist of two horizontal, parallel straight lines. Figs. 5.8a and 5.9a display double-logarithmic plots of MSD versus time for model II. The best values of d in Table 5.1 confirm the expectation.

Diffusion in the rigid microporous medium of model I bears close resemblance to diffusion in lattice gases, where, depending on concentration, the MC MSD exhibits power-law time dependences of the type t^d $(0.5 < d \leq 1)$ [225] Such processes are often called multifractal [226] when d takes more than one value. It is argued in the case of higher-dimensional lattices in which diffusion is highly restricted in all but one dimension, one would expect the MSD to exhibit a power-law dependence of the type t^d, with d about $1/2$ at intermediate times, and to become linear in t at long times. For model I, depending on α, diffusion is effectively restricted more in some dimensions than in others, so one might expect, on grounds of this argument, to observe similar diffusive behavior.

Finally, as pointed out by Feder [223], in the vicinity of a second order phase transition density correlation functions develop a component without intrinsic length or time scale. Consequently, the relevant thermodynamic potential has a critical part with scaling properties as in eq. (4.21). According to Feder [223] eq. (4.21) also implies a power law time-dependence for the MSD so that the results in this section are consistent with those presented in Sect. 4.3.2.

5.3.3 In-Plane Anisotropy of Atomic Motion

Atomic motions in x- and y-directions are inequivalent for α intermediate between 0.0 and 0.5. This is reflected in the curves of Figs. 5.8b and 5.8c, from which one may infer that motions in the y-direction is more hindered than in x-direction. This in-plane anisotropy (cf. Fig. 4.9) is slight and decreases as α approaches 0.5. For a given α it also seems to decrease with increasing time. In the MSD's obtained by MC no such in-plane anisotropy can be discerned above statistical noise.

This may be understood from the plots in Fig. 5.10 which show VACF's in x- and y-directions at $\alpha = 0.3$ as functions of time. The plots give clear evidence of in-plane anisotropy of atomic motion. They also indicate that oscillatory motion occurs predominantly in x-direction. Most important, however, in-plane anisotropy occurs on the short-time part of the physical timescale which is not amenable to MC. At intermediate times MC and MD deviate with respect to in-plane anisotropy because the short-time behavior of the latter is propagated to longer times. This can be seen from eqs. (5.5a) and (5.5b) which can be combined so that the MSD at time t is related to an integral over the VACF up to that same time. In other words,

120

the MSD at time t incorporates the short-time dynamics up to that point.

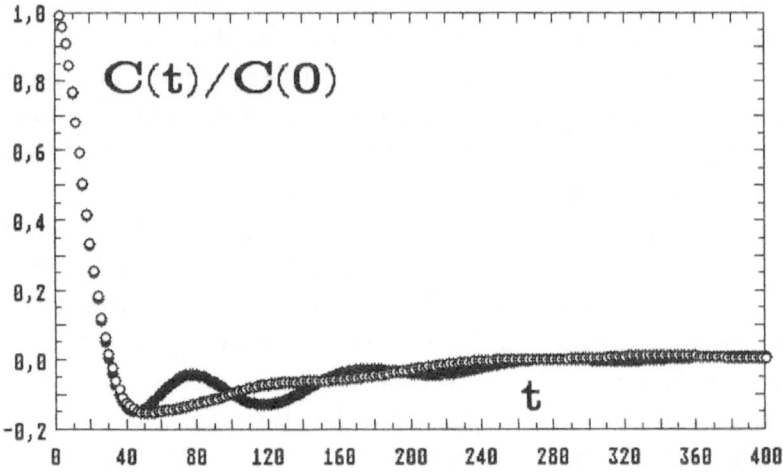

Fig. 5.10. Normalized velocity autocorrelation function versus time (units of 10^{-14} s) for a monolayer vicinal phase at $T^* = 1.00$, $-\tau_3^* = 0.234$, $N = N_s = 512$, $s^* = 25.576$, $h^* = 1.923$ and $\alpha = 0.30$; (●): x-direction, (○): y-direction.

5.4 Transport Phenomena and Non-Equilibrium Molecular Dynamics Calculations

Besides EMD NEMD provides an alternate route by which transport phenomena in fluids can be investigated [41]. Perhaps the earliest application of NEMD to vicinal phases is presented by Hannon et al. [227] who modelled the flow of dense Lennard-Jones and soft-sphere fluids between stationary walls. In their model the flow is initiated and maintained by stochastic boundary conditions at the walls similar to those suggested by Tenenbaum et al. [228]. For Reynold numbers pertaining to the laminar flow regime, Hannon et al. obtain velocity and temperature profiles in good agreement with those predicted by hydrodynamic equations for incompressible fluids [227].

The agreement holds up to pressure gradients as large as $10^4 \, \mathrm{atm \, cm^{-1}}$.

Bitsanis et al. employ NEMD to study shear viscosity and flow behavior of vicinal phases [229]. They obtain results for shear stress and effective viscosity significantly smaller than in corresponding homogeneous phases. The authors also propose the so-called "local average density model" to predict shear viscosity and diffusivity in vicinal phases. In the local average density model local transport coefficients are introduced which are taken as those of a corresponding bulk fluid at some effective mean density. The mean density is obtained by averaging the inhomogeneous density of the vicinal phase locally over a molecular volume. Results of the local average density model are in good agreement with NEMD results also presented in [229]. Bitsanis et al. [229] also obtain good agreement with theoretical predictions of Vanderlick and Davis [230] as far as diffusivity is concerned. These latter authors generalized Enskog's theory [9,10] for strongly inhomogeneous fluids.

Heinbuch and Fischer study (pressure gradient driven) Hagen-Poiseuille flow of Lennard-Jones fluids in cylindrical pores with structured walls [231]. They find different flow patterns depending on the strength of the particle-wall interaction, the strength of the gravitational-type driving force, the thermodynamic state and the mechanism of heat dissipation (i.e. the thermostat) employed at the walls. Heinbuch and Fischer's results are similar to those of Koplik et al. [232].

Hagen-Poiseuille flow of liquidlike drops through narrow cylindrical pores is investigated by Sokolowski [233]. The drops are formed during capillary condensation of Lennard-Jones molecules. The time evolution of the system depends on details of the simulation conditions as observed in Heinbuch and Fischer's work [231]. For weakly adsorbing walls whole drops slip along the pore and changes in their local structure are detected. As the particle-wall interaction increases in strength and at larger driving forces the drop tends to evaporate.

Bitsanis et al. also study (shear driven) Couette flow in narrow pores [234]. The strongly varying local density across the pore renders inappropriate the usual dependence of viscosity on local density. Bitsanis et al. [234] find a dramatic increase of the effective viscosity due to the inability of fluid layers to undergo the gliding motion of planar flow. This effect is partially responsible for strong viscosity increases observed experimentally [189] in thin films that still maintain fluidity. Bitsanis et al. [234] employ the NEMD reservoir method proposed by Ashurst and Hoover [235].

Somers et al. [236] employ MD and grand canonical ensemble MC to investigate structure and dynamics of vicinal phases at rest (i.e. in equilibrium) and under Couette flow. The authors employ structured and unstructured walls similar to models I and II. Various properties like local density profiles, pair correlation functions, diffusivity, normal pressure across the vicinal phase, the stress tensor and velocity profiles are studied in detail. Somers et al. find that for $h^* > 2.5$ the enhanced order in the contact layer (cf. Fig. 7(e)) is not altered under flow. For $h^* < 2.5$, however, that structure is significantly affected. The fluid is also sensitive to shear-induced changes in diffusivity.

Mo and Rosenberger [237] investigate fluid flow in two-dimensional channels with atomically rough walls. They employ both sinusoidal and randomly roughened walls in the spirit of model IV in this article.

Zhu and Robinson [238] simulate shock waves passing through fluids by means of an NEMD method in which vicinal phases are subjected to adiabatically moved walls. The walls are moved periodically at a certain frequency and with varying amplitude. This approach allows one to study time-dependent changes in local density. It might therefore be useful to investigate related phenomena such as solution dynamics of very large solute molecules or chemical reactions in the frozen solvent limit.

Moving walls are also employed by Lupkowski and van Swol [208] in their NEMD method. However, these authors solve the equation of motion for the walls under constant normal pressure by assigning a mass to the walls. This mass is completely arbitrary but, as pointed out by Lupkowski and van Swol, affects only dynamic properties of vicinal phases and relaxation times of equilibrium properties.

6 Summary

In this article properties and behavior of model vicinal phases are investigated computationally. It is demonstrated that the confining walls affect vicinal phases significantly in various ways. Structurally vicinal phases are distinguished from bulk phases by inhomogeneity on account of layering. Layering is effected by the confining walls which may be viewed as a representation of an external field. The inhomogeneous portion of vicinal phases dominates the structure of thin vicinal phases. This effect decreases with increasing wall separation (i. e. thickness of the vicinal phase). However, layering persists in the immediate vicinity of the walls regardless of thickness of the vicinal phase.

Details of the walls are very important for vicinal phase properties. This becomes manifest in particular in the solidification of vicinal phases which is shown to occur only if the walls are structured themselves. Depending on registry solid- or liquidlike vicinal phases can be accomodated between the walls.

Varying the registry in small steps between its extremes in- and out-of-registry melting (freezing) of vicinal phases is effected. The melting (freezing) process is shear strain driven and occurs as a second order phase transition in thin vicinal phases. The order of the phase transition is inferred from lack of latent heat release. In addition density remains a continuous function of applied shear strain everywhere. However, derivatives of density and isostress-isostrain enthalpy become maximum at the shear melting point. This is demonstrated in terms of compressibility, expansivity and isostress heat capacity which tend to diverge consistently at the shear melting point. From theoretical considerations the divergence should follow a power law. Fitting the power law to the "high" strain branch of c_{τ_3} allows one to determine the shear melting point in accord with other structural and dynamical data.

The unique characteristics of shear melting are underlined by comparison with temperature-driven melting which occurs as a first order phase transition even in very thin films considered here. Preliminary results indicate, however, that the order of the solid-liquid phase transition may depend on film thickness as far as shear melting is concerned.

In general, phase transitions in vicinal phases are significantly affected by the presence of the walls. This applies also to capillary condensation which occurs as a first order phase transition involving gas- and liquidlike vicinal phases. Experimentally capillary condensation is often accompanied by hysteresis which may occur on account of metastable thermodynamic states. Metastability in computer simulations should not be described as hysteresis but simply as an artefact caused by insufficient simulations. It is of no real physical significance in computer simulations.

Confinement also affects strongly time-dependent properties of vicinal phases. Due to the presence of the walls diffusion is anisotropic in general. Diffusion normal to the walls is reduced compared with lateral diffuion. This effect is the stronger the narrower the vicinal phase.

For very narrow vicinal phases $h^* \lesssim 3$ a hydrodynamic description of self-diffusion normal to the walls fails due to the length scale involved. This is evident from an investigation of the space- and time-dependent normal diffusion coefficient which deviates from its hydrodynamically defined counterpart below that threshold.

The precise nature of the walls is also important for details of the diffusion process. Among the four models studied here model I turns out to be the physically most realistic one because it allows for friction experienced by laterally diffusing atoms. The effect of friction is incorporated in a qualitatively correct way: it is largest for particles diffusing in the contact layer and diminishes with increasing distance from the walls. Thus, diffusion is slowest in the contact layer and increases toward the middle of the vicinal phase.

Most interesting, diffusion can be non-Brownian resulting in a power-law time-dependence of the lateral MSD's. The fractional dimension d governing the power law can be related to the shear melting point; d is the smaller the closer one gets to the shear melting point. The non-Brownian character is a result of spatial hindrance in monolayer vicinal phases. Diffusion can then be described as a two-step process. The first step corresponds to oscillatory motion in cells formed by neighboring wall atoms. The cells are connected by more or less constricted tunnels through which vicinal atoms can escape their original cells in step two. The second step is a highly cooperative process perhaps involving many vicinal atoms.

Last but not least diffusion in the (x,y)-plane can be anisotropic in the vicinity of a shear melting point. It turns out that particles are more mobile in x- than in y-direction if the system is strained along the x-axis of the laboratory frame of Cartesian coordinates.

References

1. W. Heisenberg, *Gesammelte Werke Vol. CI*, Piper (München, 1984)
2. E. Fermi, *Thermodynamics*, Dover (New York, 1956)
3. S. G. Brush, *The Temperature of History. Phases of Science and Culture in the Nineteenth Century*, Burt Franklin (New York, 1977)
4. Lord Kelvin, Nature **1**, 551(1870)
5. J. Tyndall, *Heat (Considered) as a Mode of Motion*, Longmans & Green (London, 1863)
6. J. C. Maxwell, *Theory of Heat*, Longmans&Green (London, 1863)
7. L. Boltzmann, Sitz.-Ber. Akad. Wiss. Wien (II) **66**, 275(1872)
8. D. A. McQuarrie, *Statistical Mechanics*, Harper&Row (New York, 1976)
9. S. Chapman and T. S. Cowling, *The Mathematical Theory of Nonuniform Gases*, Cambridge University Press (Cambridge, 1960)
10. J. H. Ferziger and H. G. Kaper, *Mathematical Theory of Transport in Gases*, North Holland (Amsterdam, 1972)
11. M. Schroeder, *Fractals, Chaos, Power Laws*, Freeman (New York, 1991)
12. R. L. Liboff, *Introduction to the Theory of Kinetic Equations*, Krieger (Huntington, 1979)
13. J. Loschmidt, Sitz.-Ber. Akad. Wiss. Wien(II) **73**, 128(1876); ibid. 366 (1876)
14. E. Zermelo, Ann. Phys. **59**, 793(1896)
15. E. Fick and G. Sauermann, *Quantenstatistik dynamischer Prozesse Vol. I*, Harri Deutsch (Frankfurt, 1983)
16. P. Debye, Ann. Phys. **39**, 787(1912)
17. L. Onsager, Phys. Rev. **65**, 117(1944)
18. R. J. Baxter, *Exactly Solved Models in Statistical Mechanics*, Academic Press (London, 1990)
19. J. G. Kirkwood, J. Chem. Phys. **3**, 300(1935)

128

20. B. J. Alder, Phys. Rev. Letters **12**, 317(1964)

21. A. Rahman, Phys. Rev. Letters **12**, 575(1964)

22. A. Rahman, Phys. Rev. **136**, A405(1964)

23. B. J. Alder and W. E. Alley, Physics Today, January 1984, 56(1984)

24. B. J. Alder and T. E. Wainwright, J. Chem. Phys. **33**, 1439(1960); Phys. Rev. **127**, 359(1962)

25. M. P. Allen and D. J. Tildesley, *Computer Simulation of Liquids*, Clarendon (Oxford, 1987)

26. C. Truesdell, *Essays in the History of Mechanics*, Springer (New York, 1968)

27. J. S. Rowlinson and B. Widom, *Molecular Theory of Capillarity*, Clarendon (Oxford, 1982)

28. F. M. Etzler and W. Drost-Hansen, in: *Advances in Chemistry Series*, No. 188, ed. by M. Blank (American Chemical Society, Washington D. C.) p. 486

29. J. Clifford, in: *Water: A Comprehensive Treatment*, ed. by F. Franks Vol. 5, Plenum (New York, 1975), p. 173

30. W Kauzmann, Adv. Protein Chem. **14**, 1(1959)

31. P. F. Low, Langmuir **3**, 181(1987)

32. D. H. Everett and F. S. Stone (eds.), *The Structure and Properties of Porous Materials*, Butterworths (London, 1958)

33. J. E. Shigley, *Mechanical Engineering Design*, McGraw-Hill (New York, 1977)

34. P. Laszlo, Science **235**, 1473(1987)

35. G. R. Gray and H. C. H. Darley, *Composition and Properties of Oil Well Drilling Fluids*, Gulf (Tokyo, 1980)

36. D. Nicholson and N. Parsonage, *Computer Simulation and the Statistical Mechanics of Adsorption*, Academic Press (London, 1982)

37. J. Lyklema, *Fundamentals of Interface and Colloid Science Vol. I*, Academic Press (London, 1991)

38. L. D. Landau and E. M. Lifshitz, *Statistical Physics*, Pergamon Press (Oxford, 1980)

39. W. G. Hoover, *Molecular Dynamics*, Springer (Berlin, 1986)

40. W. de Raedt and A. Lagendijk, Phys. Rep. **127**, 233(1985)

41. D. J. Evans and G. P. Morriss, *Statistical Mechanics of Nonequilibrium Liquids*, Academic Press (London, 1990)

42. H. J. C. Berendsen, in: *Proceedings of the International School of Physics 'Enrico Fermi'*, ed. by G. Ciccotti and W. G. Hoover, 1987

43. R. Vogelsang, M. Schoen and C. Hoheisel, Comput. Phys. Commun., **30**, 235(1983)

44. M. Schoen, Comput. Phys. Commun. **52**, 175(1989)

45. R. W. Hockney and J. W. Eastwood, *Computer Simulation Using Particles*, McGraw-Hill (New York, 1981)

46. D. M. Heyes and W. Smith, CCP5 Quarterly **26**, 68(1987)

47. G. S. Grest, B. Dünweg and K. Kremer, Comput. Phys. Commun. **55**, 269(1989)

48. J. Boris, J. Comput. Phys. **66**, 1(1986)

49. D. C. Rapaport, Comput. Phys. Rep. **9**, 1(1988)

50. Z. A. Rycerz, Comput. Phys. Commun. **60**, 297(1990)

51. D. J. Diestler, M. Schoen, A. W. Hertzner and J. H. Cushman, *J. Chem. Phys.* **95**, 5432(1991)

52. J. P. Hansen and I. R. McDonald, *Theory of Simple Liquids (2nd ed.)*, Academic Press (London, 1986)

53. S. A. Rice and P. Gray, *The Statistical Mechanics of Simple Liquids*, Wiley (New York, 1965)

54. T. L. Hill, *Introduction to Statistical Thermodynamics*, Dover New York, 1986)

55. E. Schrödinger, *Statistical Thermodynamics*, Dover (New York, 1989)

56. D. J. Adams, Mol. Phys. **29**, 307(1975)

57. J.P. Valleau, D.J. Diestler, J.H. Cushman, M. Schoen, A.W. Hertzner and M. E. Riley, J. Chem. Phys. **95**, 6194(1991)

58. W. W. Wood, in: *Physics of Simple Liquids*, ed. by H.N.V. Temperley, J.S. Rowlinson and G.S. Rushbrooke, North Holland (Amsterdam, 1968)

59. K. Binder and D.W. Heermann, *Monte Carlo Simulation In Statistical Physics*, Springer (Berlin, 1988)

60. N. Metropolis, A. W. Rosenbluth, M. N. Rosenbluth, A. H. Teller and E. Teller, J. Chem. Phys. **21**, 1087(1953)

61. H. Callen, *Thermodynamics*, Wiley (New York, 1966)

62. D. C. Wallace, *Thermodynamics of Crystals*, Wiley (New York,1972)

63. J. H. Weiner, *Statistical Mechanics of Elasticity*, Wiley (New York, 1983)

64. M. Schoen, D. J. Diestler and J. H. Cushman, Phys. Rev. B, 1993, in press

65. K. E. Larsson, Phys. Chem. Liq. **12**, 273(1983)

130

66. M. Nielsen, J. P. McTague and L. Russell, in: *Phase Transitions in Surface Films*, ed. by J. G. Dash and J. Ruvalds, NATO-ASI **51**, Plenum (New York, 1979)

67. J. N. Israelachvili and G. E. Adams, J. Chem. Soc. Faraday II **74**, 975(1978)

68. D. C. Prieve and N. A. Frej, Langmuir **6**, 396(1990)

69. P. K. Hansma, V. B. Elings, O. Marti and C. E. Bracker, Science **242**, 209(1988)

70. J. N. Israelachvili, *Intermolecular & Surface Forces (2nd ed.)*, Academic Press (London, 1992)

71. B. V. Derjaguin, Kolloid Zeits. **69**, 155(1934)

72. M. L. Gee, P. M. McGuiggan, J. N. Israelachvili and A. Homola, J. Chem. Phys. **93**, 1895(1990)

73. J. L. Finney, in: *Water and Aequous Solutions*, ed. by G. W. Neilson and J. E. Enderby, Adam Hilger (Bristol, 1986)

74. J.S. Clegg, in: *Biophysics of Water*, ed. by. F. Franks and S. Mathias, Wiley (Chichester, 1982)

75. W. Drost-Hansen, in: *Chemistry of the Cell Interface B*, ed. by H. D. Brown, Academic Press (New York, 1971)

76. W. Drost-Hansen, Phys. Chem. Liq. **7**, 243(1978)

77. D. C. Prieve, S. G. Bike and N. A. Frej, Faraday Discuss. Chem. Soc. **90**, 209(1990)

78. D. Rugar and P. Hansma, Physics Today, October 23-30 (1990)

79. W. A. Ducker, T. J. Senden and R. M. Pashley, Nature **353**, 239 (1991)

80. U. Heinbuch, *Fortschritt-Berichte VDI Ser. 3* **173**, VDI-Verlag (Düsseldorf, 1989)

81. G. C. Maitland, M. Rigby, E. B. Smith and W. A. Wakeham, *Intermolecular Forces: Their Origin and Determination*, Clarendon (Oxford, 1981)

82. M. Schoen, D. J. Diestler and J. H. Cushman, J. Chem. Phys. **87**, 5464(1987)

83. I. Ravina and P. F. Low, Clays and Clay Miner. **20**, 109(1972)

84. J.J. Magda, M. Tirrell and H.T. Davis, J. Chem. Phys. **83**, 1888(1985)

85. I. K. Snook and W. van Megen, J. Chem. Phys. **72**, 2907(1980)

86. D. Ruelle, *Statistical Mechanics*, Benjamin (Reading, 1983)

87. C. L. Rhykerd, Jr., M. Schoen, D. J. Diestler and J. H. Cushman, Nature **330**, 461(1987)

88. J. E. Lane and T. H. Spurling, Aust. J. Chem. **74**, 2103(1976)

89. S. Toxvaerd, J. Chem. Phys. **74**, 1998(1981)

90. G. S. Heffelfinger, Z. Tan, K. E. Gubbins, U. Marini Bettolo Marconi and F. van Swol, Molec. Simul. **2**, 393(1989)

91. S. Sokolowski and J. Fischer, Mol. Phys. **71**, 393(1990)

92. R. D. Groot, Mol. Phys. **60**, 45(1987)

93. T. F. Meister and D. M. Kroll, Phys. Rev. A**31**, 4155(1985)

94. W. van Megen and I. K. Snook, J. Chem. Phys. **73**, 4656(1980)

95. W. van Megen and I. K. Snook, Mol. Phys. **39**, 1043(1980)

96. G. M. Torrie and J. P. Valleau, J. Chem. Phys. **73**, 5807(1980)

97. S. H. Lee, J. C. Rasaiah and J. B. Hubbard, J. Chem. Phys. **85**, 5232(1986)

98. S. H. Lee, J. C. Rasaiah and J. B. Hubbard, J. Chem. Phys. **86**, 2383(1987)

99. S. de Leeuw and J. W. Perram, Mol. Phys. **66,** 637(1989)

100. K.-K. Han, J. H. Cushman and D. J. Diestler, J. Chem. Phys. **96**, 7687(1992)

101. B. Jönsson, Chem. Phys. Lett. **82**, 520(1981)

102. N. I. Christou, J. S. Whitehouse, D. Nicholson and N. G. Parsonage, Faraday Symp. Chem. Soc. **16**, 139(1981)

103. M. Marchesi, Chem. Phys. Lett. **97**, 224(1983)

104. N. Anastasiou, D. Fincham and K. Singer, J. Chem. Soc. Faraday II, **79**, 1639(1983)

105. R. Sonnenschein and K. Heinzinger, Chem. Phys. Lett. **102**, 550(1983)

106. D. J. Mulla, P. F. Low, J. H. Cushman and D. J. Diestler, J. Colloid Interface Sci. **100**, 576(1984)

107. C. Y. Lee, J. A. McCammon and P. J. Rossky, J. Chem. Phys. **80**, 4448(1984)

108. G. Barabino, C. Gavotti and M. Marchesi, *Chem. Phys. Lett.* **104**, 478(1984)

109. G. Aloisi, R. Guidelli, R. A. Jackson, S. M. Clark and P. Barnes, J. Electroanal. Chem. **206**, 131(1986)

110. E. Spohr and K. Heinzinger, J. Chem. Phys. **84**, 2304(1986)

111. E. Spohr and K. Heinzinger, Chem. Phys. Lett. **123**, 218(1986)

112. J. P. Valleau and A. A. Gardner, J. Chem. Phys. **86**, 4162(1987)

113. A. Giovanni, M. L. Foresti and R. Guidelli, J. Chem. Phys. **91**, 5592 (1989)

114. A. Delville, Langmuir **7**, 547(1991)

115. O. Matsuoka, E. Clementi and M. Yoshimine, J. Chem. Phys. **64**, 1352(1976)

116. M. J. S. Dewar and W. Thiel, J. Am. Chem. Soc. **99**, 4899(1977)

117. J. D. Honeycutt and D. Thirumalai, J. Chem. Phys. **90**, 4542(1989)

132

118. J. Fischer and M. Methfessel, Phys. Rev. A**22**, 2836(1980)

119. J. D. Weeks, D. Chandler and H. C. Andersen, J. Chem. Phys. **54**, 5237(1971)

120. J. A. Barker and D. Henderson, J. Chem. Phys. **51**, 635(1967)

121. N. F. Carnahan and K. E. Starling, J. Chem. Phys. **51**, 635(1969)

122. J. Fischer and U. Heinbuch, J. Chem. Phys. **88**, 1909(1987)

123. U. Heinbuch and J. Fischer, in: *Proc. Sec. Conf. "Fundamentals of Adsorption"*, Santa Barbara, USA, 1986, p.245

124. S. Sokolowski and J. Fischer, J. Chem. Phys. **93**, 6787(1990)

125. U. Heinbuch and J. Fischer, Chem. Phys. Lett. **135**, 587(1987)

126. S. Sokolowski and J. Fischer, Mol. Phys. **70**, 1097(1990)

127. Y. Zhou and G. Stell, Mol. Phys. **66**, 767(1989)

128. Y. Zhou and G. Stell, Mol. Phys. **66**, 791(1989)

129. Y. Zhou and G. Stell, Mol. Phys. **68**, 1265(1989)

130. G. M. Torrie, A. Perera and G. N. Patey, Mol. Phys. **67**, 1337(1989)

131. R. Evans, Adv. Phys. **28**, 143(1979)

132. P. C. Ball and R. Evans, Langmuir **5**, 714(1989)

133. P. C. Ball and R. Evans, Mol. Phys. **63**, 159(1988)

134. P. C. Ball and R. Evans, J. Chem. Phys. **89,** 4412(1988)

135. B. K. Peterson, K. E. Gubbins, G. S. Heffelfinger, U. Marini Bettolo Marconi and F. van Swol, J. Chem. Phys. **88**, 6487(1988)

136. E. Kozak and S. Sokolowski, J. Chem. Soc. Faraday. II, 1992, preprint

137. E. Kierlik and M. L. Rosinberg, Phys. Rev. A**42**, 3382(1990)

138. W. A. Curtin and N. W. Ashcroft, Phys. Rev. A**32**, 2909(1985)

139. P. Tarazona, Mol. Phys. **48**, 81(1984)

140. P. Tarazona, Phys. Rev. A**31**, 2672(1985); ibid. **32**, 3148(1985)

141. P. Tarazona, U. Marini Bettolo Marconi and R. Evans, Mol. Phys. **60**, 573(1988)

142. S. Sokolowski and J. Fischer, Mol. Phys. **70**, 1097(1990)

143. S. Sokolowski and J. Fischer, Mol. Phys. **71**, 393(1990)

144. Z. Tang, L. E. Scriven and H. T. Davis, J. Chem. Phys. **96**, 4639(1992)

145. D. Levesque and J.-J. Weis, J. Stat. Phys. **40**, 29(1985)

146. G. Rickayzen and M. J. Grimson, J. Chem. Soc. Faraday Trans. II **78**, 893(1982)

147. J. H. de Boer in [32]

148. S. J. Gregg and K. S. W. Sing, *Adsorption, Surface Area and Porosity*, Academic Press (New York, 1982)

149. E. A. Flood (ed.), *The Solid-Gas Interface Vol. 2*, Marcel Dekker (New York, 1967)

150. M. M. Dubinin, B. P. Bering, V. V. Serpinski and B. N. Vasilev, in: *Surface Phenomena in Chemistry and Biology*, ed. by J. F. Daniel, K. G. A. Pankhurst and A. C. Riddiford, Pergamon (Oxford, 1958)

151. C. H. Amberg, D. H. Everett, L. Ruiter and F. W. Smith, in: *Surface Activity Vol. 2*, ed. by J. H. Schulman, Butterworth (London, 1957)

152. D. D. Awschalom, J. Warnock and M. W. Shafer, Phys. Rev. Lett. **57**, 1607(1986)

153. G. Mason, J. Colloid Interface Sci. **88**, 36(1982); Proc. Roy. Soc. London Ser. A **390**, 47(1983)

154. R. Evans and P. Tarazona, Phys. Rev. Lett. **52**, 557(1984)

155. R. Evans, U. Marini Bettolo Marconi, Chem. Phys. Lett. **114**, 415(1985)

156. R. Evans, U. Marini Bettolo Marconi and P. Tarazona, J. Chem. Phys. **84**, 2376(1986)

157. R. Evans, U. Marini Bettolo Marconi and P. Tarazona, J. Chem. Soc. Faraday Trans. II **82**, 1763(1986)

158. B. K. Peterson, J. P. R. B. Walton and K. E. Gubbins, J. Chem. Soc. Faraday Trans. II **82**, 1789(1986)

159. P. C. Ball and R. Evans, Europhys. Lett. **4**, 715(1987)

160. J. P. R. B. Walton and N. Quirke, Chem. Phys. Lett. **129**, 382(1986)

161. W. F. Saam and M. W. Cole, Phys. Rev. B**11**, 1086(1975)

162. T. Hill, *Statistical Mechanics*, McGraw-Hill (New York, 1956), p. 164

163. J. R. Henderson and F. van Swol, Mol. Phys. **51**, 991(1984)

164. J. P. R. B. Walton, J. S. Rowlinson and D. J. Tildesley, Mol. Phys. **48**, 1357(1983)

165. J. R. Henderson and J. S. Rowlinson, J. Phys. Chem. **88**, 6484(1984)

166. J. E. Lane and T. H. Spurling, Aust. J. Chem. **33**, 231(1980)

167. W. van Megen and I. K. Snook, Mol. Phys. **54**, 741(1985)

168. D. W. Hawley, J. M. D. McElroy, J. C. Hajduk and X. B. Reed, Chem. Phys. Lett. **117**, 154(1985)

169. B. K. Peterson and K. E. Gubbins, Mol. Phys. **62**, 215(1987)

170. J. P. R. B. Walton and N. Quirke, Molec. Simul. **2**, 361(1989)

171. M. Schoen, C. L. Rhykerd, Jr., J. H. Cushman and D. J. Diestler, Mol. Phys. **66**, 1171(1989)

172. A. Z. Panagiotopoulos, Mol. Phys. **62**, 701(1987)

173. N. Quirke, Fluid Phase Equilib. **29**, 283(1986)

174. T. Poston and I. Stuart, *Catastrophe Theory and Its Applications*, Pitman (1978)

175. G. S. Heffelfinger, F. van Swol and K. E. Gubbins, Mol. Phys. **61**, 1381(1987)

176. A. de Keizer, T. Michalski and G. H. Findenegg, Pure and Appl. Chem. **63**, 1495(1991)

177. F. van Swol and J. R. Henderson, J. Chem. Soc. Faraday Trans. II **82**, 1685(1986)

178. J. W. Cahn, J. Chem. Phys. **66**, 3667(1977)

179. J. R. Henderson and F. van Swol, Mol. Phys. **56**, 1313(1985)

180. M. S. Wertheim, J. Chem. Phys. **65**, 2377(1976)

181. S. Sokolowski and J. Fischer, Phys. Rev. A**41**, 6866(1990)

182. R. Evans and P. Tarazona, Phys. Rev. A**28**, 1864(1983)

183. P. Tarazona and R. Evans, Mol. Phys. **48**, 847(1983)

184. E. Bruno, C. Caccamo and P. Tarazona, Phys. Rev. A**34**, 2513(1986)

185. J. E. Finn and P. A. Monson, Mol. Phys. **65**, 1345(1988)

186. J. H. Sikkenk, J. O. Indeku, J. M. J. van Leeuwen and E. O. Vossnack, Phys. Rev. Lett. **59**, 98(1987)

187. J. Israelachvili, P. McGuiggan, M. Gee, A. Homola, M. Robbins and P. Thompson, J. Phys.: Condens Matter **2**, SA89(1990)

188. A. M. Homola, J. N. Israelachvili, M. L. Gee and P. M. McGuiggan, J. Tribology **111**, 675(1989)

189. J. Van Alsten and S. Granick, Phys. Rev. Lett. **61**, 2570(1988)

190. J. N. Israelachvili, P. M. McGuiggan and A. Homola, Science **240**, 189(1988)

191. P. M. McGuiggan and J. N. Israelachvili, Chem. Phys. Lett. **149**, 469 (1988)

192. S. Granick, Science **253**, 1374(1991)

193. P. A. Thompson and M. O. Robbins, Science **250**, 792(1990)

194. W. Loose, *Beiträge zur Nichtgleichgewichts-Molekulardynamik-Computersimulation in einfachen Fluiden*, Dissertation (Technische Universität Berlin, 1990)

195. S. Hess, Int. J. Thermophys. **6**, 657(1985)

196. P. A. Thompson and M. O. Robbins, Phys. Rev. A**41**, 6830(1990)

197. M. Schoen, C. L. Rhykerd, Jr., D. J. Diestler and J. H. Cushman, Science **245**, 1223(1989)

198. W. W. Wood, J. Chem. Phys. **52**, 729(1970)

199. P. N. Vorontsov-Vel'Yaminov, A. M. El'y-Ashevich, L. A. Morgenshtern, V. P. Chakovskikh, High Temp (USSR) **8**, 261(1970)

200. J. G. Dash and J. Ruvalds, *Phase Transitions in Surface Films*, Plenum (New York, 1980)

201. M. Schoen, D. J. Diestler and J. H. Cushman, Mol. Phys., 1993, in press

202. H. E. Stanley, *Introduction To Phase Transitions And Critical Phenomena*, Clarendon (Oxford, 1971)

203. A. Hankey and H. E. Stanley, Phys. Rev. **B6**, 3515(1972)

204. R. B. Griffith and J. C. Wheeler, Phys. Rev. **A2**, 1047(1970)

205. T.S. Chang, A. Hankey and H.E. Stanley, Phys. Rev. **B8**, 346(1973)

206. M. Schoen, unpublished data

207. P. G. Slade, P. A. Stone and E. W. Radoslovitch, Clays and Clay Miner. **33**, 51(1985)

208. M. Lupkowski and F. van Swol, J. Chem. Phys. **95**, 1995(1991)

209. M. Schoen, D. J. Diestler and J. H. Cushman, 1993, manuscript in preparation

210. R. Kubo, M.Toda and N.Hashitsume, *Statistical Physics II (2nd ed.)*, Springer (Berlin, 1991)

211. W. A. Steele, in: *Transport Phenomena in Fluids*, ed. by H. J. M. Hanley, Marcel Dekker (New York, 1969)

212. C. Hoheisel, *Theoretical Treatment of Liquids and Liquid Mixtures*, Elsevier (Amsterdam, 1993), in press

213. J. H. Cushman, D. J. Diestler and M. Schoen, J. Am. Soc. Mech. Eng. Div. Appl. Mech. **117**, 17(1991)

214. R. Zwanzig, Phys. Rev. **133**, A50(1960)

215. J. P. Boon and S. Yip, *Molecular Hydrodynamics*, McGraw-Hill (New York, 1980)

216. G. Arfken, *Mathematical Methods For Physicists*, Academic Press (London, 1985), p. 450ff

217. P. L. Hall and D. K. Ross, Mol. Phys. **36**, 1549(1978)

218. I. N. Bronstein, K. A. Semendjajew, *Taschenbuch der Mathematik (25th ed.)*, B. G. Teubner (Stuttgart, 1991)

219. M. Schoen, J. H. Cushman, D. J. Diestler and C. L. Rhykerd, Jr., J. Chem. Phys. **88**, 1394(1987)

220. R. Zwanzig, J. Chem. Phys. **33**, 1338(1960)

221. B. J. Berne and G. D. Harp, Adv. Chem. Phys. **17**, 63(1970)

222. A. Einstein, Ann. Phys. **17**, 549(1905)

223. J. Feder, *Fractals*, Plenum (New York, 1988)

224. J. H. Cushman, M. Schoen and D. J. Diestler, to be published, 1993

225. K. W. Kehr and K. Binder, in: *Applications of the Monte Carlo Method in Statistical Physics*, ed. by K. Binder, Springer (Berlin, 1987)

226. B. B. Mandelbrot and J. W. Van Ness, SIAM Rev. **10**, 422(1968)

227. L. Hannon, G. C. Lie and E. Clementi, *Phys. Lett.* **A119**, 174(1986)

228. A. Tenenbaum, G. Ciccotti and R. Gallico, Phys. Rev. *A25*, 2778, (1982)

229. J. Bitsanis, J. J. Magda, M. Tirrell and H. T. Davis, J. Chem. Phys. **87**, 1733(1987)

230. T. K. Vanderlick and H. T. Davis, J. Chem. Phys. **87**, 1791(1987)

231. U. Heinbuch and J. Fischer, Phys. Rev. **A40**, 1144(1989)

232. J. Koplik, J. Banavar and J. Willemsen, Phys. Fluids **A1**, 781(1989)

233. S. Sokolowski, Phys. Rev. **A44**, 3732(1991)

234. J. Bitsanis, S. A. Somers, H.T. Davis and M. Tirrell, J. Chem. Phys. **93**, 3427(1990)

235. W. G. Hoover and W. T. Ashurst, Adv. Theor. Chem. **1**, 1(1975)

236. S. A. Somers and H. T. Davis, J. Chem. Phys. **96**, 5389(1992)

237. G. Mo, F. Rosenberger, Phys. Rev. **A42**, 4688(1990)

238. S.-B. Zhu and G. W. Robinson, Chem. Phys. **134**, 1(1989)

Lecture Notes in Physics

For information about Vols. 1–379
please contact your bookseller or Springer-Verlag

New Series m: Monographs